GENERATIVE AI, MEDIA, AND SOCIETY

This groundbreaking book demystifies generative AI's transformative impact on media, socio-cultural dynamics, ethics, and policy.

Defining generative AI as an evolutionary leap in the development of artificial intelligence, the author examines intricate human-machine interactions and socio-technical dynamics, advocating robust, proactive AI governance to address emerging uncertainties. The book is clearly structured into six key chapters, each exploring distinct aspects of the relationship between artificial intelligence, media, and society. The chapter on "Transformation" examines how machine behavior is reshaping our datafied society, questioning whether data is the new oil, or digital manure. "Generative AI" investigates the models and future impacts of generative AI as a co-intelligence, revisiting the Turing Test and analyzing societal-business impacts. "AI Media" explores the convergence of media and AI, highlighting robot journalism, synthetic content, and the disinformation era and discussing the trend toward high-risk optimism. "Uncertainties" addresses inherent unpredictability vs. strategic foresight, focusing on challenged business models, sustainability concerns, and emotional intelligence factors. "Ethics" analyzes generative morality and dual-use technology, covering trusted AI principles—from misuse to integrative solutions. Finally, "Policy" discusses governance, labor market impacts, and the importance of human rights and power dynamics in generative AI. Each chapter also provides summaries of impact projects, reflective art, scholarly questions, and strategic takeaways—extended with a comprehensive glossary.

This is an essential resource for scholars, students, policymakers, technologists, ethicists, and AI industry leaders seeking to rapidly understand and address the challenges and opportunities of generative AI and AI media in a cohesive framework.

Katalin Feher is a globally recognized expert in generative AI, AI media, and socio-technical systems. With extensive research, policy, and industry experience across four continents, her work focuses on responsible AI governance, societal impacts, and strategic foresight.

GENERATIVE AI, MEDIA, AND SOCIETY

Katalin Feher

Routledge
Taylor & Francis Group
LONDON AND NEW YORK

Designed cover image: © Katalin Feher, the author

First published 2025
by Routledge
4 Park Square, Milton Park, Abingdon, Oxon OX14 4RN

and by Routledge
605 Third Avenue, New York, NY 10158

Routledge is an imprint of the Taylor & Francis Group, an informa business

British Library Cataloguing-in-Publication Data
A catalogue record for this book is available from the British Library

ISBN: 978-1-032-96875-9 (hbk)
ISBN: 978-1-032-96873-5 (pbk)
ISBN: 978-1-003-59102-3 (ebk)

DOI: 10.4324/9781003591023

Typeset in Sabon
by Taylor & Francis Books

To Alex, Luca, and Lujza

CONTENTS

THE AUTHOR

Katalin Feher's expertise spans generative AI, AI in media, socio-technical systems, and responsible-explainable AI. She is a public intellectual with extensive research, policy, and industry experience across four continents, including significant work in North America, Europe, Southeast Asia, the Far East, the Middle East, and East Africa.

She is a transatlantic Horizon Europe NGI Professor, affiliated with the Ludovika University of Public Service as an Associate Professor in Hungary and collaborating with Concordia University's Applied AI Institute in Canada. As an external expert, she serves the United Nations' Global Digital Cooperation, the European Innovation Council EISMEA, European Commission MSCA, and the Future of Rights and Governance (FORGE) at New York University. She is Co-Chair of the Mediatization Section of the European Communication Research and Education Association and an editorial board member at the *KOME* journal. Feher is an invited member of the Media As SocioTechnical Systems (MASTS) research community at the University of Southern California, and the Coalition for Independent Technology Research in the U.S. She is the Founder and Principal Investigator of AI Media Research, the first independent global content hub in the emerging R&D&I of AI and media. Recently, she was a Fulbright Research Scholar at Drexel University in the United States and a Bolyai PostDoc at the Hungarian Academy of Sciences. She holds a Ph.D. in Communication and Media Studies, having completed a habilitation in Philosophy. Her publications are available in D and Q-ranked social science and multidisciplinary academic journals. She is also a regularly invited keynote speaker in North America and Europe at prominent academic, business, and policy conferences or high-level symposiums.

Katalin Feher values time with her loving family, including her supportive partner and wonderful daughters. She is also a travel junkie, kickboxer, runner, and contemporary art photographer.

The manuscript was prepared and written in Aarhus, Amsterdam, Boston, Brno, Brussels, Budapest, Copenhagen, Istanbul, Ljubljana, London, Los Angeles, Mombasa, Montreal, Nairobi, New York, Nicosia, Philadelphia, Riga, Rome, Toronto, Vancouver, Velence, Washington D.C., and Zurich between 2022 and 2024.

FUNDS AND ACKNOWLEDGMENTS

European Union's Horizon Europe Research and Innovation Programme – NGI Enrichers, Next Generation Internet Transatlantic Fellowship (Grant number: 101070125) and the Janos Bolyai Research Scholarship of the Hungarian Academy of Sciences supported the book (Grant number BO/00045/19/9).

This monograph has been shaped by the experience, inspiration, and insights gained through combined research, business, and policy collaborations, leading global AI conferences and contributions from high-level expert groups, including United Nations Global Digital Cooperation OSOPs, EmTech conferences and networks at MIT, FORGE at New York University, responsible AI research at Concordia University's Applied AI Institute, socio-technical research at the University of Southern California MASTS, the Fulbright research in AI-driven media at Drexel University, policy in research integrity at the Coalition at Independent Technology Research community, collaborations by Informa research, the leading narratives of European Communication Research and Education Association (ECREA), Ministry for Innovation and Technology (NRDI)-funded AI research at the Corvinus University of Budapest, the community of LSE Media Futures, the MIT program in Policy for Science and Technology, and the most significant global AI community, Inspired Minds, with balanced networks in business, policy, and academia.

I extend my deepest gratitude to several other communities whose collaborative spirit and invaluable insights have been the cornerstone of this book's creation. I sincerely thank my fellow researchers, participants, and engaged Ph.D. students in related projects for their endless enthusiasm in the journey that led to this milestone. I am grateful to my research, policy, and corporate colleagues across my extensive networks for their invaluable ideas and shared content, which inspired my work. Special thanks to the innovators who broadened my perspective and colleagues whose sharp critiques pushed me forward, even those who offered brief but encouraging feedback.

Thank you to the Department for Science Strategy, the Department of Social Communication, and the Science and Society Research Group at the Ludovika University of Public Service in Hungary for encouraging the process and supporting the experience.

Sincere thanks are extended to the case study providers for their invaluable insights, especially AI @ The FBI, the European Digital Media Observatory, and the Data School at Utrecht University.

Gratitude is extended to the reviewers for their constructive feedback, as their comments have been instrumental in refining the substance of this work.

The entire publishing team's exceptionally professional, high-quality, and efficient work advancing the monograph's publication is greatly appreciated.

Acknowledgment is extended to my husband, whose insightful discussions on technology consistently challenge and enrich my academic thinking. Our young daughters, deeply fascinated by emerging technology, not only grasp its potential but have also started to understand the boundaries we set. Their perspective has also shaped this book, written with a profound sense of responsibility toward the impact on future generations.

PREFACE

Setting the Stage

The emergence of generative AI has been a gradual evolution, gaining momentum well before its most recognized applications surfaced. As academic research and discourse around AI expanded significantly, so did the imperative to scrutinize its transformative effects on society. While discussions surrounding AI's impact on media, ethics, and governance have been growing, the explosive rise of generative AI—epitomized by tools like ChatGPT—has pushed these debates into the center of public discourse. Its rapid adoption highlighted this technology's potential and significant challenges. Alongside this advancement, the proliferation of mis/disinformation exposed the challenge and a need for a structured, multidisciplinary analysis to navigate the complexities of generative AI's impact.

The Core Objective

This monograph provides a comprehensive summary of the rapidly emerging field of generative AI and its intersection with AI media. It is intended to demystify these technologies and study their specific implications with future impacts. As generative AI evolves, it becomes crucial to distinguish it from broader AI concepts because generative AI focuses explicitly on creating original content, designs, and insights based on existing data patterns. This capability disrupts traditional notions of creativity, authorship, and authenticity, transforming how we understand and interact with media and information systems. This work studies the ethical dilemmas, uncertainties, and regulatory challenges that generative AI and AI in media present, emphasizing the role of adequate policymaking, human-centered governance, and strategic foresight in addressing these complexities.

Who Benefits Most?

This book is crafted for those committed to understanding the transformative power of generative AI and synthetic media in society, including academics, researchers, policymakers, industry leaders, practitioners, ethicists, tech enthusiasts, digital activists, and students in related studies across various disciplines. The monograph offers a synthesized understanding of critical concepts, issues, cases, and artworks, making it an essential resource for professors, graduate students, Ph.D. candidates, and professionals in Media Studies, Sociology, Political Science, Computer Science, Ethics, Policy Studies, Business, and Art Studies. It also provides valuable insights for experts in Law, Information Studies, Public Administration, Environmental Studies, Decision Science, and Cultural Studies. Moreover, it addresses business and policy fields such as data privacy, AI governance, intellectual property, market and foresight research, and sustainability development while relevant to enforcement agencies and policymakers in AI regulation. Tech innovators, corporate leaders, and stakeholders in Creative Industries will benefit from its interdisciplinary approach, which will help shape GenAI policies and foster cross-industry collaboration. Artists and creative industry representatives will find opportunities and risks to explore, learn new terminology, and encounter groundbreaking exploratory artworks. This volume, therefore, is a comprehensive guide to exploring the complexities of generative AI and its impact on media and society.

INTRODUCTION

> I am a techno-neutralist. I seek to uncover how synthetic worlds consume our physical and digital ones and how we face this intertwined generative AI reality.
>
> —*The author*

Over the last few years, I have been caught up in a whirlwind of debates and predictions at top AI industrial conferences. The atmosphere was exciting as keynote speakers took the stage, from business and governmental representatives to professors, policymakers, and faces from law enforcement agencies. On the surface, everything appeared clear and optimistic, with polished presentations emphasizing concepts such as *responsible AI* and *explainable AI*, albeit sometimes blurring the lines between AI and generative AI. However, behind the scenes, in the speakers' room, discussions were marked by uncertainty and genuine concerns, especially in the emergence of generative AI. Even more striking was the sight of big tech executives, flanked by security, slipping out of sessions without addressing the urgent questions raised by scientists and policymakers.

At prominent academic conferences, I have faced another dimension of this evolving story. Researchers and scientific communities, whose work has long been critical in shaping the ethical framework for AI, have expressed frustration that while the demand for their expertise in responsible and transparent AI was skyrocketing their institutional funding and resources were being reduced. At the same time, governmental representatives were asking them to do more with less, and inviting them to fix the problems of generative AI and disinformation technologies.

In parallel with all this, I've observed at firsthand how generative AI tools have quietly started integrating into routine processes within high-level policymaking circles, subtly influencing decision-making and operational workflows. There was always a satisfied smile when they shared how generative AI had helped them craft more engaging speeches or improve their work efficiency. However,

DOI: 10.4324/9781003591023-1

the smiles would disappear when the conversation turned to the societal impact of AI. Even when they make significant efforts to develop tools and establish expert boards in this field, there is a palpable unease about the growing challenges, the erosion of public trust, and the impact on democratic systems.

Generative AI represents an evolutionary leap in the development of artificial intelligence. It introduces capabilities that expand its potential in creative industries, personalization, and human-machine collaboration. While academic literature primarily addresses technosolutionism, socio-technical imaginaries, and democracy challenges (Schaake 2024; Verdegem 2021, Lindgren 2013), the generative milestone requires a broader perspective of human-machine interplay. It is essential to recognize that now we may transform our reality in a manner similar to that in which we transformed spoken language into its written form. However, generative AI brings a fundamentally different shift in how we encode and interpret reality, as it reinterprets existing global human knowledge and synthesizes diverse information and content based on complex calculations.

When synthetic elements begin to devour non-synthetic ones—which is imminent—significant consequences are likely to follow. We have both the responsibility and the opportunity to shape this merged synthetic reality in the following decades and establish guiding principles for its development. As foundational models in generative AI serve as building blocks for innovation, so the foundational principles provide a framework for foundational concepts of ethics, governance, and future challenges in an AI-driven world.

This monograph, therefore, examines the evolving phenomena of generative AI and synthetic outputs, demystifying their impacts and highlighting enduring trends amid rapid technological advancement. The book is designed, thus, for a diverse audience; it invites policymakers, industry professionals, academics, tech enthusiasts, AI ethicists, and university students to engage with relevant topics through a comprehensive summary. Ultimately, this monograph equips readers with informed, concept-driven strategies for shaping generative and synthetic futures. It explores transformative technology's profound influence on society, economy, and culture, offering a comprehensive framework to understanding and addressing these shifts. The following chapter structure guides this exploration.

- *TRANSFORMATION.* This section explores transformative technologies and datafication, their societal impact, AI evolution, socio-technical systems, and machine behavior.
- *GENERATIVE AI.* This segment analyzes generative AI's origins, mechanics and models, and societal-business impacts from entertainment to NewsTech—with implications for the future.
- *AI MEDIA.* This section explores AI's transformative impact on media, from robot journalism to synthetic immersion, while addressing mis/disinformation, manipulation, and misplaced optimism.

- *UNCERTAINTIES*. This chapter explores uncertainties in business models, socio-environmental sustainability, emotional intelligence, and societal perceptions.
- *ETHICS*. This section examines machine agent behavior, balancing human values, mitigating bias, harm, or colonialism, and promoting responsible, explainable AI practices.
- *POLICY*. This chapter explores governing generative AI, focusing on human rights, labor, power dynamics, and strategies for harnessing technology for social innovation.

Accordingly, readers will develop a comprehensive understanding of generative AI's expanding role in media, business, and governance while addressing societal challenges shaping the technology's future.

Each section includes four key insights beyond the core content: impact projects that showcase innovative applications, reflective artworks that interpret discussed challenges, scholarly questions designed to support scientific research and university learning, and strategic takeaways offering a practical summary for academia, industry, and policymaking.

The final sections, featuring a comprehensive glossary and index, offer hands-on knowledge for readers to quickly grasp key terminology and emerging buzzwords, enhancing their competence in a field that will have a long-term, profound societal impact.

Happy (non-generative?) reading!

1

TRANSFORMATION

Defining Transformation

Transformation involves the profound, ongoing reconfiguration of societies, industries, and individuals, driven by integrating advanced technologies such as AI, quantum computing, and algorithmic processes. This shift alters daily interactions and decision-making and restructures system architectures, offering new opportunities while also introducing complex challenges. In a datafied society—where data serves as oil and manure—decision-making, governance, and personal behaviors are increasingly transformed by data-driven systems. Autonomous technologies and robotics, embedded within socio-technical frameworks, co-evolve with human behavior, fundamentally altering labor markets, communication, and social roles. These transformations extend to broader societal constructs, influencing knowledge production, human autonomy, and the nature of technological relationships. As transformation advances, it necessitates critically examining intelligence, agency, and the evolving human-technology relationship.

Introduction

Mentioning artificial intelligence (AI) instantly sparks a mix of curiosity, concern, and speculation among listeners, regardless of their background. The universal reaction underscores AI's pervasive intrigue and the myriad questions it raises about its current capabilities, future potential, and ethical implications. This book's opening chapter aims to demystify transformation technology in society by interpreting its essential terminology, and the key areas it influences. Given the diverse interpretations of AI—from machine learning to broader technological suites—this foundation clarifies AI's scope and explores its significant role in transformation.

DOI: 10.4324/9781003591023-2

Digital transformation integrates computational and networked technologies into all aspects of business or society, fundamentally altering how they operate and deliver value to users. AI plays a central role in this shift by simulating and augmenting human intelligence processes, including learning, reasoning, and decision-making. AI operates unobtrusively behind the scenes, enhancing efficiency, productivity, creativity and personalization without direct human intervention, driving innovation across various sectors as a *silent intelligence*. The evolution of generative AI, highlighted by models like GPT, has redefined content creation, making it possible to produce human-like text or multimodal content generation (Ferrara 2024). This technological advancement, a dramatic anthropological shift, originates from the development of binary code by Gottfried Wilhelm Leibniz centuries ago, which enabled the representation of complex information using only 0s and 1s; it is also the foundation for modern computing and digital systems, allowing machines to process information efficiently (Strickland & Lewis 2022). Future computing paradigms like *quantum computing* could alter this binary logic by using *quantum* bits (qubits), representing 0 and 1 simultaneously, potentially transforming how information is processed (Nielsen & Chuang 2010).

This chapter summarizes transformative technologies and their comprehensive impact on society, focusing on key areas. Emerging concepts of datafication, deep tech, AI winters and springs, black box technologies, open-source ecosystems, and technosolutionism underscore the complexities and challenges of adopting transformative innovations. The chapter also highlights surveillance, autonomy, and socio-technical error, reflecting on debates around human intelligence, post/transhumanism, and the nature of digital transformation in shaping the future.

Transformation Technology

According to Lewin (1947), transformation refers to the process of changes at the system level leading to alterations in its fundamental properties or structures. It often requires transitioning from an existing state to a desired future state. Christensen (2015) posited that transformation technology pertains to innovations that disrupt and reshape existing industries and societal structures. More recently, Brynjolfsson and McAfee (2014) also expanded on this, suggesting that rapid advancements in digital technologies exemplify transformative forces in modern economies.

Transformation technology in media, information, communication, and entertainment industries is driving a convergence of mediums, blurring the lines between digital and physical experiences and creating new platforms for storytelling. Manuel Castells (1996) theorized that transformations in these sectors involve shifts in how societies produce, distribute, and consume content in the face of technological advancements. A study by Jenkins (2006) resonated with all these via the concept of the convergence culture, where old and new media intersect, reshaping the entertainment and information dissemination structures.

In the 21st century, the convergence of human cognition and machine intelligence emerges as a pivotal theme in the evolution of technology. Aue's (2018) human-to-machine communication model epitomizes this trend, underscoring that for machines to augment human cognitive capabilities, they must grasp the quintet of human thought: perception, motivation, reasoning, learning, and knowledge. This model promises a symbiotic future where machines comprehend and emulate human cognitive processes, fostering an unprecedented collaboration between neural and silicon-based intelligence.

Simultaneously, quantum computing represents a paradigm shift in hardware capabilities, potentially accelerating AI applications. Unlike classical bits, quantum computing leverages quantum mechanics principles, using qubits that can exist in superposition. This enables faster problem-solving in areas like cryptography and optimization. *Quantum encryption* is also integral to AI evolution and deployment, enhancing data security by making unauthorized access instantly detectable.

Though still experimental, it has the potential for transformative breakthroughs across industries. The fusion of quantum computing with AI, particularly in media and generative AI applications, signifies a transformative shift in content personalization and security protocols. When these quantum advancements interlace with AI's cognitive models, it promises a future of unprecedented processing capabilities, highlighting an AI-quantum convergence that could redefine the boundaries of machine learning, data analysis, and the very nature of computational creativity (Biamonte et al. 2017).

This convergence is not merely a technological leap; it's a catalyst for a synergy of human-like artificial cognition, and the computational power of quantum systems could unlock mysteries in complex systems and redesign industries. With this convergence, we edge closer to a world where the interplay between quantum computing and AI could lead to advancements once thought to reside solely in science fiction.

Zeineddine (2020) optimistically suggests that this fusion could solve global challenges, such as poverty and disease. Yet, this optimism is tempered by the pragmatic realization of our planet's finite resources, posing the question: Does the marriage of human and artificial intelligence hold the key to sustainable problem-solving, or will it exacerbate the race toward resource depletion?

Accordingly, a resource surge should be channeled toward technological advancements by crafting policies that make them available. Likewise, Mona Sloane (2019) reflects on the socio-political implications of AI, urging us to consider the philosophical and policy frameworks that will shape our future, calling for a careful orchestration of innovation and regulation.

The discourse of technological transformation is replete with instances of AI's expanding capabilities. Prasad (2020) champions the need for AI systems to eliminate bias and ensure fairness while possessing broad and deep knowledge, efficient learning, and the ability to reason for decision-making. This reflects an ambition to create machines that process information and understand and engage with the world ethically and effectively.

Yet, this transformation is not without its critics. The birth of Xenobots, as reported by Sanjeev (2023a), represents a milestone where AI-designed organisms are brought to life using biological materials. This blurs the lines between the creator and the creation, questioning the very essence of life and evolution. Are we, as a species, ready to coexist with entities whose genesis lies within artificial algorithms?

While some view technology as a tool, others like Faraj and Leonardi (2022) argue for a relational reconceptualization, where technology is perceived as a dynamic web of evolving relationships deeply embedded in organizational strategy and practice. This shift from a tool-centric to a relation-centric view of technology necessitates a profound transformation in strategizing and innovating.

Deep tech refers to innovations based on cutting-edge scientific and engineering advancements that address complex global challenges (De la Tour et al. 2017). Related technologies such as AI, super apps, genetic algorithms, quantum, edge computing, 5G/6G networks, and blockchain are rapidly transforming industries. These innovations drive an increasing demand for data, often called data hunger (Steinhoff 2022). While this surge in data fuels technological progress, it also raises concerns about privacy, energy consumption, and ethical use. By examining these technologies through the lens of data hunger, we can better understand their societal impact and question how the relentless pursuit of data may transform social constructions and sustainability.

Technosolutionism, as critiqued by Britt Paris et al. (2023), cautions against the over-reliance on technological fixes for social issues. They advocate for an Internet infrastructure that incorporates an ethics of care, challenging the dominance of libertarian and market-driven ideologies. It's a call for reimagining technology as part of a broader societal fabric, responsive to the needs of communities rather than the whims of market forces. In contrast, the technological imperative, a concept that suggests the inevitability of technological progress, often leads to the assumption that society must adapt to technology rather than vice versa. However, Christodoulou and Iordanou (2021) argue for a more balanced approach, where democratic processes are strengthened to ensure that technology serves the public good rather than dictating the terms of its engagement.

In the backdrop of these philosophical and ethical considerations, human beings witness tangible threats such as the Black Mamba malware, a sophisticated and evasive cyber threat targeting critical systems, a stark reminder of the dark side of technological progress. Balaji (2023) highlights the need for robust cybersecurity measures to combat such sophisticated threats that have the potential to undermine the very fabric of AI trust. Although transformative technologies may not initially attract immediate adoption, instructional resources such as YouTube videos, internal training programs, and clear guidelines can help users overcome resistance and unlock a powerful pathway to transformation.

As we ponder such transformations, the black swan concept looms large. Black swans, as defined by Nassim Nicholas Taleb (2005), are rare and unpredictable events that have a massive impact, often beyond our expectations or control, fundamentally altering industries or societal structures. Unforeseen consequences, like those arising from the creation of Xenobots, artificial life forms created by combining robotics and biology (Spichak 2021), force us to confront the uncertainty inherent in this transformative journey. These unpredictable developments highlight how such disruptive events can reshape our understanding of technology and its broader implications, posing significant risks and opportunities that demand careful consideration and preparedness.

In conclusion, transformative technology is not merely about transmuting bits and atoms but about transforming the narrative of human existence. It's a story of harnessing AI to expand human potential, interpret ontological paradigms, and critically examine technosolutionism. Ultimately, transformative technology seeks to ensure that, as we move into the future, narratives of responsibility and collective visions guide us toward a sustainable and equitable world.

Data as Oil or Manure

The debate about whether data becomes the oil or the manure of the digital transformation is discussed in different forums. It provokes visions of a future where AI's role is as ubiquitous as energy in the industrial age. The analogy of data as oil suggests a fuel that powers the engines of innovation, a precious commodity that drives the artificial intelligence economy. Conversely, likening data to manure proposes a more organic view, where data enriches the fertile fields of technology, allowing for cultivating diverse AI applications (Feher 2016b).

Initiatives to enable its self-driving cars to self-diagnose and order parts autonomously, as Lutz (2023) describes, crystallize this discussion. It represents the *autonomy* AI might achieve, raising significant questions about the extent of control and trust we place in these systems. What unforeseen capabilities might sprout when such autonomous systems interact or, more intriguingly, collide with each other by accident or design?

Considering this example, AI's expansive toolbox, ranging from Natural Language Generation to Emotion Recognition, produces different ways of human-machine co-decisions, blurring the boundaries between models and tools. The autonomy of these tools and models, their ability to act without human supervision or even belief, is increasingly a subject of public and business policies. Consider, for example, the intriguing potential of reading the brain through MRI data, where the inner visualizations of a person can be reconstructed with startling clarity. This development is always twofold, from opportunities for exploratory research to privacy concerns.

In this data-driven society, a *data lakehouse* emerges as a repository combining a data lake's vast storage capacity with a data warehouse's management capabilities. This hybrid structure aims to maintain data sovereignty and governance while hastening the transition from concept to production.

The evolution of humanoid robots, from smart appliances to androids like Sophia, exemplifies the tangible manifestation of AI. These embodiments of AI present the promise of companionship and service but also raise questions about social integration and the replication of complex human behaviors, such as the subtleties of facial expressions and the nuances of cultural recognition.

Gehl and Lawson (2022) touch upon the darker side of this data-driven reality with social engineering. They analyze the manipulation of communication, highlighting the new forms of influence and deception that emerge as data becomes a tool for fraud. In line with this, Klenk (2024) urges conceptual engineering to design values and criteria for a non-manipulative generative AI.

In psychological operations, the role of large language models (LLMs) in shaping social narratives cannot be understated, especially given the popular use of generative AI. Fredheim (2023) warns of AI's vulnerabilities, pointing out that it is susceptible to targeted prompts that could be exploited to manipulate public opinion or reveal sensitive information.

In this context, is data the oil that lubricates and propels the advancement of AI, or is it the manure that, while potentially messy and odorous, enriches the ground for a diverse and sustainable growth of intelligent systems? The answer may lie in how we harness this resource—whether we drain it for a short-term surge of power or cultivate it to yield a rich harvest of innovation and wisdom for future generations.

Datafied Society

The transformation of the World Wide Web from its nascent stage to Web 2.0, and its ongoing evolution into the semantic Web 3.0, marks a pivotal shift in the fabric of our digital society. This metamorphosis defines a *datafied society*, a term that encapsulates the profound integration of data into every facet of human existence with ubiquitous interconnectivity. Central to this shift is the paradigm of human-machine collaboration. This symbiotic alliance melds the innate cognitive prowess of humans with AI's computational might to forge outcomes unattainable by either entity in isolation or partly connected.

In this data-saturated construction, humans assume the pivotal role of "data donors" for AI, their interactions and information serving as the bedrock upon which intelligent systems are sculpted. AI platforms, burgeoning from this fertile ground, have emerged as the cornerstone of myriad applications and services, weaving their influence through the fabric of diverse sectors. The advent of edge computing signifies a strategic pivot, bringing data processing closer to its genesis, thereby enhancing efficiency and paving the way for innovations in real-time applications. In generative AI, a "prompt" refers to the user-provided input or instruction that guides the AI to generate specific outputs, including text, images, music, code, models, and designs. Prompt engineering, as an emerging discipline, focuses on crafting and refining such prompts to optimize AI responses, thus increasing precision and effectiveness across diverse tasks (Megahed et al. 2024). This approach enhances human-AI collaboration by enabling more targeted interactions, ultimately improving the integration of AI in various applications.

Beneath the surface of technological innovation lies the critical need for infrastructural thinking, emphasizing the importance of robust, scalable systems to support the weight of an ever-expanding digital ecosystem. This foundational perspective is essential for unlocking the transformative potential of AI and automation, transforming industries, enhancing efficiency, and creating novel business models. However, this progress is fraught with complex challenges, requiring us to address large-scale risks and confront the moral and ethical dilemmas tied to such profound technological changes.

Polanyi's paradox posits that knowledge and understanding are demonstrated not only in verbal expression but in human intuition as well. Highlighting the tacit, often indescribable knowledge unique to human cognition, Polanyi's paradox serves as a crucial reference point in the design of AI systems. It pushes us to create technologies that aspire to replicate the subtlety and nuance of human decision-making while confronting the limitations inherent in automation and machine learning. Conversely, the emerging *Trust Paradox* reveals a dual challenge: We struggle to trust knowledge from sources we do not understand while simultaneously failing to comprehend what we do not trust (Zhang et al. 2024). This dynamic may restrict the effective use of AI-generated insights, as a lack of understanding can undermine the trust needed for their application.

The narrative of digitalization highlights the infrastructures driving global connectivity, where platforms and technologies continuously evolve. As 5G networks roll out, they reveal stark disparities across geographies shaped by policy, economic forces, and innovation, as noted by Carroll (2023). At the same time, the rise of Web 3.0—driven by blockchain and NFTs—challenges centralized platforms, signaling a future filled with innovative, decentralized digital services (Galloway 2022). 5G offers up to 10 Gbps speeds and low latency, supporting IoT, autonomous vehicles, and VR/AR. Meanwhile, 6G, expected by 2030, will deliver speeds of up to 1 Tbps, integrate AI, and enable futuristic applications like holograms and brain-computer interfaces, heralding a more augmented experience in AI.

As AI morphs into an on-demand service, we stand on the cusp of a paradigmatic shift that transcends mere productivity enhancements, venturing into the redefinition of work and value creation. This transition, however, is fraught with ethical and existential dilemmas, where the imperative to mitigate extreme risks becomes paramount. The discourse on "infrastructure-thinking," as expounded by Hjarvard (2022), unveils the socio-political implications of infrastructural dominion, where the monopolization by tech titans raises alarms about potential censorship and the undermining of democratic tenets.

In this intricate dance of datafication, a peculiar paradox emerges—the ouroboros of AI's evolution, a self-referential loop where AI feasts on its synthetic progeny, distorting the tapestry of authenticity. This "model collapse," highlighted by Shumailov et al. (2024), is a stark reminder of the delicate equilibrium between human oversight and the autonomous burgeon of technology. Standing at the crossroads of infrastructure and platformization (Mansell & Steinmueller 2020), a narrative woven from the threads of power, control, and innovation delineates the contours of our digital existence. In this epoch of datafication, the ouroboros not only symbolizes the inherent potential for self-destruction but also epitomizes the cyclical nature of progress, where each terminus heralds a new genesis, and every commencement portends the possibility of divergent denouements.

In a datafied society, where data is central to decision-making and governance, responsible dataset construction is essential. Orr and Crawford (2024) highlight the significance of this conceptual framework, identifying key challenges in *dataset construction*, such as balancing data scale and quality, managing resource constraints, addressing biases introduced by shortcuts, and confronting the consequences of ambivalence over accountability.

This approach emphasizes that datasets are not neutral entities but rather shaped by the social, technical, and organizational contexts in which they are constructed. The framework introduced by Orr and Crawford is helpful because it encourages a critical examination of the underlying factors influencing dataset construction, such as biases, resource limitations, and accountability issues. By fostering this critical perspective, their work promotes a deeper understanding of how datasets directly impact fairness and ethical AI practices. However, while this approach is crucial in recognizing the complexities involved in data creation, it could benefit from more actionable strategies for addressing these challenges in practice. Identifying the issues does not necessarily ensure solutions, especially in environments with significant resource constraints or institutional inertia. Nevertheless, it is a valuable starting point for advancing responsible data use in a data-driven world.

Definition and Four Waves of AI

Artificial intelligence refers to the simulation and enhancement of human intelligence processes by machines, particularly computer systems, encompassing learning, reasoning, and behavior. Its evolution has occurred in four waves so far: the initial phase of rule-based systems, followed by adaptive learning models, then AI capable of interacting within complex environments, and finally, the emergence of generative AI, which integrates human and machine outputs, transforming our understanding of existence and future interactions.

The *first AI Wave* of handcrafted knowledge started when our aspirations were as constrained as our technologies, where AI systems relied on rules meticulously etched by human hands. As Tzezana (2017) notes, this era was defined by its structured, logical, and often brittle systems, which could perform well-defined tasks under specific conditions but faltered outside their crafted confines.

The second AI Wave brought the winds of change with statistical learning. The systems became more fluid and adaptable, leveraging statistical models to learn from large datasets. This wave was characterized by its probabilistic nature, a shift from the rules' rigidity to the patterns' flexibility. During this time, AI began to touch the lives of the masses, subtly weaving its way into the fabric of daily life with recommendation systems and optimized searches.

Yet, it is the *third AI Wave, contextual adaptation*, where we begin to see AI emerge as a tool and an assistant or companion of sorts. This wave is defined by systems capable of understanding and adapting to complex, dynamic environments—going beyond learning patterns to grasp the nuances of context and meaning. This wave, still nascent and full of potential, brings us closer to an AI that can interact with different aspects of existence with deep augmentation of human cognition and ontology.

The fourth wave of AI is defined by the rise in generative AI that can generate synthetic content with substantial impact, profoundly transforming human cognition, knowledge production, creativity, narratives, and predictive capabilities. Its fusion of human and machine outputs raises critical questions about human agency and the fragility of truth and facts, and emphasizes the need for frameworks for engaging its societal implications. Moreover, the fourth wave is also challenging or transforming the results of the first three waves of the AI paradigm. The related critical inquiries will be explored deeply in the following chapters to untangle the complexities of generative AI's influence.

AI winters and springs are changing accordingly. AI winter refers to a period in AI research and development characterized by reduced funding, waning interest, and skepticism about the field's potential. This is primarily due to overly optimistic expectations that could not be met with the existing technology and understanding. The first significant AI winter occurred in the 1970s, followed by others decade by decade. These winters resulted from the realization that early AI algorithms and hardware could not achieve the ambitious goals initially set out. Conversely, an AI spring describes periods of renewed enthusiasm, investment, and progress in AI. During these times, breakthroughs in theory, methodology, and computing power reinvigorated the field, leading to significant advancements and applications. AI springs are characterized by a surge in funding from both public and private sectors, an increase in published research, and the commercialization of AI technologies. The current era, marked by advancements in deep learning, natural language processing, and machine learning, is an ongoing AI spring driven by unprecedented computational power and large datasets.

However, what binds these waves together is the notion of trust, a trust that stems from understanding. Explainable AI (XAI) becomes the beacon in this quest for clarity. As posited by DARPA, a research and development agency of the United States Department of Defense (2018), XAI seeks to create "glass box" models, offering transparency and interpretability to the processes and decisions of AI systems. It allows a "human-in-the-loop," ensuring that AI systems can be trusted and their actions understood, bridging the gap between capability and comprehension.

With each wave, the types of AI have diversified from analytic and functional to interactive and visual, as Bekker (2019) categorizes. Yet, the looming question persists: Are we approaching the end of the distinction between narrow, general, and super AI? Will these categories meld into one another, or will they become even more pronounced as AI evolves?

As Karnouskos (2022) explores, the implications of these waves extend into law and society. The symbiosis between humans and AI has ethical conundrums and legal challenges. How do we assign liability when AI systems operate autonomously? Will the robot ever be seen as a tool and a legal subject? And as AI systems become more integrated into our lives, how do we ensure they do not encroach upon our liberties, even as they seek to protect us? Tons of societal questions are floating around. The third wave especially pushes AI closer to the nebulous boundaries of cognition and consciousness. It echoes with questions about what it means to be intelligent, learn, and adapt. As McCarthy et al. (2006) remind us, it is a time when AI begins to mirror the evolutionary processes that have shaped life on this planet. The intertwining of ecological and evolutionary principles with AI development suggests a future where AI may not just be a phase but a fundamental aspect of life's continual adaptation.

In conclusion, the Four Waves of AI represent the ebb and flow of our technological aspirations. From the handcrafted knowledge to the generative turn, AI has evolved to reflect our growing understanding of artificial and organic intelligence.

Machine Behavior

Machine behavior refers to how autonomous systems—such as bots, robots, and intelligent agents—act within their environments, making decisions and taking actions based on their programming and goals. These systems are designed to interact with both humans and other machines, often exhibiting complex behaviors that can impact social, economic, and political dynamics. Machine behavior involves understanding the technical capabilities of these systems and the consequences of their actions in various contexts, such as ethical, social, and political implications.

Notable examples include *political bots* used in *computational propaganda*, as Woolley and Howard (2016) and Howard and Kollanyi (2016) described. By amplifying certain messages or spreading mis/disinformation through social media, these bots manipulate public opinion. Exhibiting behaviors that blur the line between legitimate public discourse and AI manipulation, these bots disseminate information, often without transparency. Such actions raise significant concerns about their role in shaping public debates.

Intelligent agents, like the aforementioned bots, also demonstrate increasingly autonomous behaviors. While optimizing processes, they can exacerbate surveillance and privacy violations. As these agents operate within data-driven societies, their behaviors influence how individuals interact with information, contributing to the attention economy and often overwhelming users with content overload and information overload, as highlighted by Wu (2018).

Machine behavior, as explored by Rahwan et al. (2019), examines how AI systems interact within human environments through interdisciplinary studies spanning AI, cognitive science, sociology, and media studies. In the context of Generative AI and AI-driven media, machine behavior becomes particularly relevant in shaping human-AI interactions. Generative AI, such as AI systems that autonomously generate text or media content, influences how information is created and disseminated, raising concerns about authenticity, bias, and mis/disinformation. AI-driven media also shapes the attention economy, as algorithms increasingly curate and flood users with content. These fields demand critical examination of feedback loops, as AI learns from human input while simultaneously shaping societal behavior and discourse, raising ethical challenges around transparency and control in online ecosystems.

As Generative AI continues to evolve, recent studies like Epstein et al. (2023) explore its capacity to autonomously generate media content, raising essential questions about the ethical boundaries between human and machine-created outputs. The blurred lines between content created by humans and machines bring authenticity, trust, and mis/disinformation concerns. Generative AI models, capable of producing highly realistic text, images, and videos, challenge traditional notions of authorship and accountability, as they can be easily manipulated to deceive or mislead audiences. This is especially significant in AI-driven media, where rapid AI-generated content can contribute to information overload, complicating users' ability to distinguish between credible sources and manipulated data.

Strategic foresight in machine behavior and its interdisciplinary connections can focus on human-machine co-behavior. This involves examining how humans and machines influence one another in real time, especially within hybrid systems where AI operates alongside humans. Research could explore the feedback loops between human inputs—such as biases, emotions, and social tendencies—and how AI systems, in turn, shape those same human behaviors. *Socio-informatics* (Wulf et al. 2018), especially, can play a critical role in understanding machine behavior by examining the interplay between social structures, human behavior, and information technologies. It explores how AI systems impact society and culture, emphasizing the co-evolution of technology and social systems, where machine behavior both shapes and is shaped by human interaction and societal norms. In generative AI and AI-driven media, human trust and behavioral decision-making are essential in news filtering, social media interactions, content creation, or moderation. All these raise the question of what cooperative behaviors will emerge between human workers and AI systems in various fields.

Socio-technical Systems

In the fluid tapestry of socio-technical systems, the intertwining threads of human interaction and technological innovation form a complex weave that underpins the digital economy. The European Union's Digital Economy and Society Index (DESI) serves as a barometer, gauging the vitality and maturity of these systems across key areas such as human capital, connectivity, digital technology integration, and public services. This index is a compass pointing toward a future where technological performance and societal progress are interdependent, revealing the intricate dance of policy, practice, and people.

Polanyi's paradox is at the heart of this dance—acknowledging that much of human knowledge is tacit, understood inherently rather than explicitly. The craftsman's unspoken intuition and the artist's silent discernment defy codification. This tacit knowledge is both a challenge and a frontier for AI development, raising the question: How can we teach machines what we cannot articulate?

Yet, the transformation unfolding within our socio-technical fabric is not without its shadows. The dark alleys of tracking, profiling, and manipulation reveal societal divides and political polarization, intensified by the proliferation of mis/disinformation and hate speech. Within this context, reality-check teams strive to uphold truth, countering the distortions that threaten the very bedrock of our shared reality. Ferrara's argument about the synthetic direction of our socio-technical evolution is particularly relevant in this context, as it highlights the increasingly blurred lines between human-driven and AI-driven realities. Thus Ferrara's argument is, in brief: Deepfakes and entirely synthetic realities, powered by generative AI, challenge the very essence of reality and raise concerns about media trust and societal perception. Misusing these technologies threatens societal structures by enabling large-scale manipulation, potentially undermining democracy, social cohesion, and public trust. Therefore, robust mitigation strategies, ongoing ethical guidelines, and continuous monitoring are crucial in ensuring that these powerful technologies are developed and deployed responsibly, particularly in media and cultural-creative industries (Ferrara 2024).

The original Turing Test, proposed by Alan Turing, sought to discern if machines could think by imitating human responses. But as the socio-technical systems evolve, we face a reverse problem: Can humans discern the machinations of AI, or are we being outpaced by machines that simulate our cognitive abilities?

This socio-technical transformation is indeed profound, as Muhlhoff (2020) posits, reshaping our interactions with machines and the very power structures and societal norms that have long governed us. As AI roles transition from tools to participants in social processes, our traditional understandings are challenged, necessitating a reevaluation of ethics and the dynamics of human-AI collaboration. Within this complex framework, the socio-technical perspective provides a lens through which we can view AI as a co-creation that requires interaction between social and technical processes.

Accordingly, the socio-technical perspective recognizes the complex interplay between social and technical processes. Informed by Hayles's concept of "cognitive assemblages" (2017), this approach challenges simplistic views of technology as mere tools for human use. Hayles reveals that cognition extends beyond conscious thought, encompassing nonconscious processes essential for human and technical systems to function. These cognitive assemblages—where human and machine cognition interact—transform societal structures and reshape our understanding of agency. This perspective is particularly relevant in exploring how AI and other advanced digital systems assist and alter our social construction and narratives, driving profound changes in both conscious and unconscious dimensions of human experience.

The narratives spun by search and research technologies, such as Google's generative AI search feature, are not just algorithms in action; they are socio-technical orchestrations that shape public understanding and manage critiques within a controlled framework. This strategic engagement underscores the efforts of entities like Google to maintain their dominant position while influencing public discourse around search functionalities.

As we ponder the pause of giant AI experiments, advocated by open letters from institutes like the Future of Life Institute (2023), and as we confront AI-amplified outcomes that challenge our grasp on truth, we must acknowledge the potential of socio-technical systems to either elevate or undermine the integrity of our shared reality. *Socio-technical errors* (Leveson 2016) also arise when technological systems fail to fully account for the social complexities and behaviors of users, leading to unintended consequences, such as reinforcing biases or spreading mis/disinformation. However, these errors highlight the delicate balance required to align technological advancements with societal values, illustrating how misalignment between human needs and AI outcomes can either foster progress or deepen societal divisions.

The socio-technical systems we inhabit today are not static constructs; they are living, breathing entities that evolve with every click, every swipe, every algorithmic decision. They reflect our aspirations, our biases, and our collective will. Every choice in human-machine interaction shapes our AI-driven society, weaving a future of unprecedented benefits alongside evolving ethical challenges.

Attention Economy

The *attention economy* is both a battleground and a marketplace. Where human focus is the most sought-after commodity, data is a currency, and screens serve as stalls; businesses compete for consumers' fragmented attention (Qi 2020). As the economy increasingly revolves around capturing and sustaining attention, transformative technologies are crucial in shaping business models and societal behaviors. These technologies transform how attention is commodified and how information is consumed, controlled, and circulated.

One of the most critical aspects of the attention economy is its connection to the rise of artificial intelligence. Embedded in everything from social media algorithms to news feeds, AI systems curate content based on user preferences, learning from each interaction to better predict and manipulate what users will engage with next. As Wu (2018) argues, the economy of attention exacerbates information overload, creating a digital context where users or audiences struggle to discern meaningful content amid a deluge of distractions. The convergence of biotechnology and disruptive technologies, as Webb (2021) suggests, further intensifies this attention-grabbing dynamic by creating synthetic realities, much like mRNA vaccines enter, act, and disappear, leaving only traces in memory. This approach is an apt metaphor for how digital or AI-driven content engages users and audiences fleetingly.

In pursuing attention, transformative technologies—specifically AI—carry significant societal implications. As Ramani and Wang (2023) point out, while AI holds the potential to pivot the axis of the attention economy, it faces several constraints, such as economic productivity, technical hurdles, and societal challenges. The result is a Baumol effect in which certain parts of the attention economy thrive while less dynamic sectors anchor others.

The dark side of the attention economy is also illuminated in Pasquale's *Black Box Society* (Brevini & Pasquale 2020), where personal data becomes the foundation of algorithms-driven reputations and financial power. Here, attention is not merely a commodity. It is a mechanism of control, resulting in opaque systems governing information, decisions, and algorithms, hiding the inner workings of complex data-driven technologies. Algorithms operate within black-boxed systems, where the inner workings are hidden from users and even regulators, making it difficult to assess how decisions are made and whose interests are prioritized (Brevini & Pasquale 2020). Cohen (2019) echoes this sentiment, arguing that commodifying personal data and attention shifts power dynamics, reducing personal autonomy while enriching those who can harness these digital economies most effectively.

One stark contrast to black-box technology is the rise of the open-source ecosystem, where software is freely accessible, adaptable, and shareable. While this ecosystem encourages innovation and transparency, it has challenges. Maintaining quality control and security in decentralized environments remains a concern, particularly as unchecked changes can undermine the integrity of open-source projects (Chesbrough 2003). However, the openness of this ecosystem is a crucial counterpoint to the closed nature of black-box systems, presenting a more democratic, albeit fragile, technological construction. As a transformative technology, open-source tools have the potential to accelerate advancements across industries while encouraging collaboration and knowledge-sharing.

Another dimension to consider in the attention economy is the temporality of data centers, as described by Velkova and Plantin (2023). These data hubs are more than mere storage facilities; they dictate the rhythms of our digital lives, determining the pace at which information flows and how and where our attention is directed. In this sense, data centers act as gatekeepers of the attention economy, controlling the what and when of information dissemination. Their operations mirror the broader dynamics of power and control, where those who command the infrastructure draw, manufacture, and manipulate attention.

The attention-driven economy comes into another perspective when examining how businesses exploit this resource. As Lauer (2021) critiques, business models built around engagement frequently profit from sensationalism, extremism, and mis/disinformation. The fines levied against tech giants for exploiting attention are merely the cost of doing business in an economy where profit is tied directly to how much time and focus users can be drawn into digital platforms.

Despite these critiques, efforts exist to create an innovation-friendly ecosystem that values attention without exploiting it. For instance, the European Union's Digital Single Market aims to foster an environment that respects users' attention, promoting transparency and ethical use of data. However, the tension between optimizing attention and respecting it as a human experience remains unresolved.

Paraphrasing Maslow

Paraphrasing Maslow's hierarchy of needs offers a provocative lens to examine our technologically entwined existence. At the foundation of this reimagined hierarchy lies the "human donor for AI," feeding the data and algorithms that shape our perceptions, desires, and actions.

The stark disparities highlighted by AI penetration data challenge the structure of Maslow's classical hierarchy, introducing data-interconnectivity and technological literacy as fundamental needs. This shift questions the structure of the traditional human needs pyramid. It emphasizes the human-artificial dichotomy, juxtaposing the imperfections inherent in human nature and the natural world against the perceived precision of artificial intelligence. This reevaluation proposes a nuanced understanding of our needs, prompting a reflection on the role of technology and imperfection in shaping survival strategies.

The *half-life of knowledge*, as Charles (2019) describes, refers to the rapid rate at which professional expertise becomes outdated, a process increasingly accelerated by the advancements in artificial intelligence. In other words, as AI systems evolve and continuously transform industries, the relevance of established knowledge and skills diminishes faster than ever, creating a need for constant adaptation and learning. This phenomenon highlights the critical impact of AI on professional sectors, where staying current requires adapting to an ever-shifting labyrinth of technological innovation. As AI evolves, so must human understanding and cognitive capabilities.

Additionally, cultural and ethical diversity grapples with the globalization of AI's influence, as Gupta and Heath (2020) note. The absence of geographic diversity in AI ethics echoes a larger issue of representation and inclusion. In transformative technologies, echo chambers foster environments where algorithms selectively amplify content that aligns with users' beliefs, reinforcing biases and constraining diverse perspectives. As AI's reach extends, it must carry diverse voices and values or risk becoming an echo chamber of the few rather than a chorus of the many.

Transhumanism, which advocates for enhancing human capabilities through technology, redefines self-actualization by pushing beyond the limits of human potential. It envisions a future where individuals transcend biological constraints using advanced technologies such as AI, genetic engineering, or cybernetic enhancements.

In contrast, *posthumanism* explores the dissolution of traditional human boundaries, focusing on how technology fundamentally alters human identity and existence, potentially leading to new forms of being that transcend current notions of humanity altogether. These concepts challenge existing hierarchies of achievement, questioning the essence of what it means to be human. Both concepts represent a transformative technology-based society that reinterprets Maslow's hierarchy of needs, suggesting that human needs, such as self-actualization, can be met not only through traditional means but by exceeding natural limitations through technological augmentation and reimagining human existence.

Yet, as we ascend this pyramid, we are confronted with the question: Can we live, work, and enjoy without these transformations, or has the promise of AI become inseparable from our conception of progress? Maslow's pyramid, in the context of AI, must now accommodate the networked connectivity that underpins our social genes—our data. This hierarchy is no longer a structure to climb but a network to maneuver, where the paths are not steps but links, and the apex is not a point but a platform.

Foundation models, as Rabowsky (2023) discusses, are not just tools; they are the bedrock of our digital Maslow's hierarchy. They underpin the synthetic content generated by AI, blurring the lines between the creator and the created. The future intelligences Papp (2023) categorized as transcendent, augmented, autonomous, and generative represent the new tiers of our needs and aspirations. Redefining "human" in a trans- and posthuman world, the implications extend beyond the personal to the societal, organizational, and geopolitical. The definition of "human" becomes fluid, expanding to include digital or synthetic entities and architected beings, challenging our preconceived notions of identity and existence.

The *Chinese room argument* posited by Coeckelbergh (2020) reminds us that despite our advancements, the nature of consciousness and the essence of human experience remains beyond the reach of machines. As we spiral through the loops of normal and abnormal behavior, where AI-driven feedback systems nudge us toward average normality, we risk losing the richness of human diversity. Additionally, Pacey's (1985) critique of *dangerous determinism*, the belief that all events—including human actions—are determined by external forces, sometimes negating free will and moral responsibility, highlights the importance of reassessing values in diverse contexts. This critique underscores the risks of attributing inevitable progress to technology without considering its impact on societal structures and individual autonomy, pushing for a deeper reflection on innovation imperatives, such as inclusivity, adaptability, sustainability, and ethical responsibility.

Paraphrasing Maslow comprehensively, we are not merely reimagining the hierarchy of needs but redefining which pyramid elements should be augmented to accommodate the emerging dynamics of human-machine co-intelligence and co-behavior. This involves a fundamental shift in understanding how machines interact with human cognition and emotional needs, suggesting that the traditional layers of human development—self-actualization—may need to integrate more complex, collaborative forms of intelligence where human and machine competencies are intertwined.

Impact Project

The Open Source Program Offices (OSPOs) under the United Nations Global Digital Cooperation initiative illustrate how AI and digital transformation technologies are harnessed for international cooperation and societal good. This initiative primarily addresses the ethical and practical challenges surrounding deploying AI and digital public goods, focusing on the public sector. By establishing OSPOs, the aim is to promote open-source solutions that are scalable, transparent, and accessible to a wide range of stakeholders, fostering global collaboration and enabling nations to adopt cutting-edge AI technologies. The long-term goal of OSPOs is to facilitate the sustainable digital and generative transformation that aligns with the UN's Sustainable Development Goals (SDGs), driving societal progress and ensuring that no one is left behind in the transformative competition.

The OSPOs initiative emerged in response to growing concerns about the barriers to digital transformation in many parts of the world, particularly in low- and middle-income countries. Despite the rapid advancements in AI and other advanced digital technologies, these communities often remain on the fringes of technological transformation, facing access, affordability, and digital and generative AI literacy challenges. In the context of AI, significant regulatory and ethical considerations exist, including the potential for deepened inequalities, privacy violations, and the misuse of AI technologies. AI-driven media tools, for example, can exacerbate mis/disinformation and widen the digital divide, raising urgent questions about governance and responsible development. By leveraging open-source technologies, the OSPOs initiative seeks to address these challenges and create a more inclusive digital ecosystem. By their nature, open-source software and AI tools offer opportunities to bypass traditional barriers like licensing fees and vendor lock-in, allowing countries to implement solutions tailored to their unique needs.

To solve these issues, the OSPOs framework provides an open, practical, and globally scalable solution. OSPOs act as designated offices within organizations that support and guide the adoption of open-source software, policies, and best practices. These offices are critical in driving innovation, creating public digital infrastructure, and enabling global cooperation. One of the key components of this initiative is the Open Source Ecosystem Enablement Framework (OSEEF), which serves as a guide for countries to adopt open-source solutions that are sustainable and impactful. The initiative also includes the Open Source Technical Facility (OSTF) and the Global Knowledge Hub for Open-Source Public Services, providing technical support and a centralized knowledge repository. These structures help local actors implement AI and digital technologies to improve public services, ranging from AI-driven healthcare to AI-enhanced disaster response systems, and they also use communication and generative systems.

The impact of the OSPOs initiative is already being seen across multiple sectors. In addition to improving public service delivery, the initiative has fostered new partnerships between governments, academia, and the private sector. This enhances international cooperation and provides tangible benefits to underserved communities. By democratizing access to AI and digital technologies, OSPOs support the development of more equitable and resilient societies, ensuring that the benefits of digital transformation are shared across borders.

In conclusion, the OSPOs initiative under the UN Global Digital Cooperation framework is a powerful example of how generative AI and open-source technologies can be leveraged to promote social good, international cooperation, and sustainable development. The initiative highlights the importance of transparency, collaboration, and inclusivity. Expanding stakeholder engagement and refining policies will be essential to ensure the continued success of this initiative.

Reflective Art

Transformative technologies such as AI and machine learning are increasingly becoming integral to creating and interpreting artwork, marking a significant shift in art and creativity. These technologies augment the artistic process and challenge traditional notions of authorship, creativity, and the artist's role. By blending computational power with human ingenuity, artists explore new frontiers of expression, producing works that reflect the complexities and possibilities via transformative and generative technologies. Through this fusion, art becomes a mirror, reflecting our technological advancements' societal, economic, and cultural implications. Three examples represent this mirroring below.

The Next Rembrandt is a project that utilized deep learning algorithms to analyze Rembrandt's 346 known paintings, mimicking extracting style, brushstrokes, and subject matter. By synthesizing this data, AI generated a new artwork, embodying the convergence of historical artistry and contemporary technology, challenging the traditional boundaries of artistic creation, and raising questions about the future role of AI in art (Brinkhof 2021).

DeepDream by Google (2024) was initially a tool to understand neural networks' inner workings. DeepDream became an artistic phenomenon, turning ordinary images into psychedelic masterpieces. This project highlighted AI's ability to reinterpret visual data, offering a glimpse into how machines perceive the world and transforming our understanding of art and creativity.

Hereafter explores the poignant intersection of memory, loss, and technology by creating *generative ghosts*—digital artifacts created from individuals' online activities and interactions, forming AI-generated personas or representations that remain after their physical departure and embody their past digital engagement and identity (Morris & Brubaker 2024). The application probes the boundaries of synthetic legacy and the human desire for immortality, prompting reflection on the evolving relationship between our digital identity or online self (Feher & Katona 2023, Feher 2021, Feher 2017, Feher 2016a, Feher 2014) and the afterlife.

These artworks created through transformative technologies serve as critical touchstones for understanding the profound social, economic, and cultural shifts precipitated by technological advancements. They exemplify the innovative use of AI and machine learning in creative expression and provoke reflection on the evolving relationship between humans and machines. By challenging conventional boundaries of art and authorship, these works illuminate the potential for technology to reinterpret creativity in society. Furthermore, they underscore the impact of digital transformation on the economy, particularly in the creative industries, and reflect on the cultural implications of living in an increasingly data-driven world. Ultimately, these artworks invite us to consider the broader implications of our technological trajectory intuitively as well.

Conclusion

Transformation is the profound and ongoing reconfiguration of society, industries, and individuals, driven by interconnected advanced technologies. These shifts affect daily interactions, decision-making, and system structures, bringing opportunities and challenges. Generative AI enhances information and content creation, producing human-like outputs (Ferrara 2024). However, it also raises concerns about authenticity, intellectual property, and the ethical implications of AI-driven content.

Datafication, as a central concept, describes data that informs decisions and transforms governance, influencing several societal aspects, from public policy to human behaviors. This evolution is most evident in socio-technical systems, where human and machine behavior interact dynamically, redefining professional knowledge and social structures (Strickland & Lewis 2022). As human-AI co-behavior becomes integral to various systems, concerns arise about the long-term impact on trust, autonomy, and societal norms. In parallel, quantum computing introduces a new paradigm for information processing by leveraging quantum bits (qubits), speeding up computational capacities (Nielsen & Chuang 2010). These developments prompt critical reflections on human autonomy, digital rights, inclusivity and the balance between innovation and control. While these technologies offer new pathways for growth, their integration into society requires interdisciplinary approaches, responsible governance, and a precise alignment with human values.

Scholarly Debates

For Cultural and Social Sciences

- How do cultural narratives about AI in media shape public understanding, societal expectations, and perceptions of technological advancements?
- How does AI-driven transformation challenge or redefine traditional notions of identity within diverse cultural and social contexts?
- How does the economy of attention transform cultural consumption patterns and social interactions in a data-driven society?
- How do transhumanism and posthumanism challenge or reinforce existing cultural and social norms in the age of AI?

For Science, Technology, and Engineering

- How do the different waves of AI development influence the design, functionality, and future potential of emerging technologies?
- How does datafication impact the technical design of AI systems and their broader societal applications?
- What technical challenges arise in developing AI systems that can contextually adapt to diverse and dynamic environments?
- How can engineers effectively mitigate the risks associated with black box technology in AI systems?

For Business, Policy, and Innovation

- How does the attention economy shape business practices, marketing strategies, and consumer behavior in a digital transformation?
- How does AI-driven transformation drive business models and strategy changes across various industries?
- What are the potential risks and benefits of technosolutionism in addressing complex societal challenges?
- How can policymakers create effective frameworks that balance the need for innovation with the protection of public interests?

Strategic Takeaways

For Academia

- Adapting Traditional Theories: Re-contextualize classic socio-cultural theories to reflect the complexities and dynamics of a datafied society, where data-driven processes increasingly shape social and cultural interactions.

- AI and Behavior: Reconsider and expand established cognition, decision-making, and social dynamics models to account for the evolving influence of AI systems and human-machine co-behavior.
- Quantum Computing Impact: Investigate the transformative effects of quantum computing on socio-technical systems, focusing on its potential to alter data processing, security, and computational infrastructure.
- AI's Socio-political Implications: Explore AI's broader societal implications, emphasizing the need for diverse, inclusive, ethical frameworks that address the socio-political challenges posed by its integration into governance, policy, and social systems.

For Industry

- AI Storytelling Innovation: Utilize AI technology to transform storytelling and content delivery in media sectors.
- Advancing AI Interfaces: Improve AI understanding of human thought for more intuitive user experiences through advanced communication models.
- Leveraging Quantum AI: Explore quantum computing to boost AI's data processing and machine learning capabilities for a competitive edge.
- Ethical AI Development: Commit to creating bias-free, fair AI systems. Embed ethical considerations from inception.

For Policymaking

- Bridging the AI Divide: Formulate policies ensuring universal access to connectivity and AI literacy as basic needs.
- Balanced AI Regulation: Develop regulations that foster AI innovation while safeguarding ethical use, particularly in cognitive enhancement.
- Partnerships in Quantum AI: Promote collaborations to advance quantum computing in AI, focusing on security and national interests.
- Strengthening Cybersecurity: Implement robust cybersecurity protocols against threats like sophisticated malware, ensuring digital ecosystem trust.

2

GENERATIVE AI

Defining Generative AI

Generative AI is a technology that produces synthetic outputs—including content, designs, and models—from learned data patterns, transforming creativity, business processes, and societal structures. Unlike traditional AI, which focuses on analyzing data and making predictions, generative AI actively creates new outputs while amplifying AI's capabilities across several industries. By generating data and improving tasks like fraud detection and personalization, generative AI pushes AI beyond its conventional limits. This interplay accelerates innovation but raises challenges like mis/disinformation, bias, and misuse, demanding ethical and regulatory frameworks to guide its transformative potential. Concepts such as the stochastic parrot question whether AI truly creates or merely mimics patterns, while issues of watermarking and poisoned data raise concerns about authenticity and manipulation.

Introduction

During the early adoption of generative AI (GenAI), there was a narrative that we would hear at academic and business conferences: "I consulted ChatGPT..." (or Bard, Gemini, Co-Pilot etc.). Critics have viewed this as diminishing originality, while supporters have appreciated the blend of human knowledge and AI, using about 10% to 20% of internet data (HubSpot & Jasper 2023). This contrast highlights GenAI's role in human-machine collaboration, with models like GPT-4 trained on diverse sources, including books, Wikipedia, and social media, totaling 570 GB or 300 billion words, a vast linguistic collection (Lockett 2023, Vinchon et al. 2023).

The challenge of originality is just one aspect of a broader dilemma. Another critical issue is the fine-tuning of these models. To curate such extensive

DOI: 10.4324/9781003591023-3

datasets, companies often depend on low-paid, non-AI professionals exposed to potentially harmful GenAI content (Perrigo 2023) to sift and structure the data, raising concerns over service quality and ethics. Large Language Models (LLMs) and other generative AI systems are increasingly shaping a global, consensual social filter bubble (paraphrasing Pariser 2011), regardless of the specific methodologies or outcomes. This has significant implications for standards in creative industries, bias assessment, and professional knowledge, potentially diminishing the richness and diversity of global digital discourse.

The popular narratives highlight GenAI as a significant milestone in artificial intelligence history, introducing groundbreaking content creation and data processing capabilities and reaching a broad audience. This development of technological evolution signals a paradigm shift in our engagement with and comprehension of digital platforms and services.

Alongside this, critics often see generative AI as simple imitation rather than true intelligence. The related current approaches are confused several times and interpret GenAI as a "bullshit machine" or "sociopath technology" that is not programmed to understand human values (Sanjeev 2023b). These critics argue that GenAI's limitations are attributed to the technology rather than the nature of language or contents, even if this can be interpreted as two sides of the same coin.

Browning and Lecun (2023) provide a more accurate interpretation, as abstract language limits AI's understanding of concrete information. Although historically AI research equated linguistic ability with knowledge, as seen in the Turing Test, this view is challenged by context-dependent and ambiguous language that restricts AI's depth of understanding. A similar argument may be valid for generative media in general, including audio and video content. This real-world approach establishes GenAI fundamentally and scientifically, strengthening confidence in the technology.

Critics argue that generative AI lacks accurate understanding, labeling it a copy without human values (Sanjeev 2023c). However, Browning and Lecun (2023) provide a balanced view, attributing GenAI's limitations to the abstract nature of language, which affects its understanding of concrete information. This challenges the idea that linguistic ability equals knowledge, highlighting language's context dependence and ambiguity. This insight extends to generative media like audio and video, grounding future GenAI applications in scientific reasoning and fostering trust in its capabilities.

Amidst these considerations and the swift progress of GenAI, its initial shortcomings are being overcome, with noticeable enhancements emerging monthly. While still debatable, the benefits and impacts of GenAI are increasingly apparent. GenAI is not merely a technological leap; it's a driver of broad societal change, redesigning content creation and transforming industries like banking, healthcare, and education. The future of generative AI narrates an entangled AI and its profound integration and influence on society and the creator economy. Therefore, this chapter suggests a comprehensive review of the GenAI phenomenon.

Origins of Generative AI

The genesis of generative AI is marked by intricate progress and unforeseen innovation. Initially conceptualized in the mid-20th century in the broad sense of AI and the Turing Test, this intricate and unforeseen genesis set the groundwork for future developments. The specific advancements are focused on machine learning and neural networks. Generative AI evolved noticeably in the 1980s and 1990s by developing more complex models. A significant breakthrough came in the mid-2010s with the creation of *Generative Adversarial Networks* (GANs), which enabled the generation of highly realistic images, text, and audio using extensive datasets. Broadly interpreted, GANs were recognized more as a media or advertising technology than a future signal of generative AI, even if Variational Autoencoders (VAEs) emerged, focusing on data encoding and generating diverse outputs. Integrating these neural network architectures with vector databases for semantic relationships and context has expanded the capabilities of generative AI, marking a substantial leap in AI innovation (Gartner Futures Lab 2023). This innovation transformed content creation by allowing AI to significantly reduce the time and effort required for these tasks. It opened up new possibilities in creative fields, allowing for the rapid prototyping of ideas, personalization at scale, and the development of entirely new forms of media that were previously unattainable through traditional methods.

The 2010s saw the rise of Large Language Models (LLMs) like GPT, culminating in the release of ChatGPT in 2022. This showcased an innovative level of human-like text generation. As Kruger (2023) highlights, this field aims to replicate and surpass human abilities. Since 2018, academic research on the socio-cultural impacts of AI has also surged (Feher & Katona 2021), laying the foundation for the current generative AI hype.

The *catalysts for transformative change* in generative AI include breakthroughs in deep learning, access to big data, advanced computational power, innovative algorithms, and increased demand for creative automation. These advancements have transformed generative AI into a tool for automating content creation to transform effectiveness and creativity, accelerating *design discovery* for prototyping, fostering creative innovation, and detecting fraud to mitigate misuse of the technology (Apablaza-Campos et al. 2024, Liao et al. 2024, Selvarajan et al. 2024). Thus, generative AI powers applications in different industries, offering innovative solutions and significantly boosting productivity and creativity.

Lockett (2023) emphasizes the critical role of extensive, varied datasets in training AI models, highlighting that this is not merely a straightforward process but a complex science of pattern learning and imitation. These datasets serve as a foundation for AI, enabling it to identify and replicate intricate patterns, ranging from language syntax to artistic nuances. Similarly, developing neural networks, mainly through deep learning, parallels an artist's journey from basic sketches to detailed artistry, deepening our understanding of the functions of the human brain.

The rise of generative models goes beyond simple text or content generation. They engage in meaningful conversations and demonstrate reasoning abilities, blurring the line between human and machine intelligence. Their skill in producing coherent, context-relevant text is notable, yet it prompts research and ethical considerations. Moreover, even as generative AI augments human intelligence with its vast, digitized knowledge, key concerns are how such a technology might eclipse human creativity or frameworks of intellectual property.

Therefore, the application of generative AI in fields ranging from art to realistic speech is both groundbreaking and controversial. Generative AI is not just a technological feat; it mirrors our human complexities, aspirations, and societal challenges. However, GenAI is not just about the technology itself; it also involves a massive amount of human labor in developing and fine-tuning generative AI services. Mir (2023) provides a clear example of how low-cost labor in data labeling and moderation raises significant moral questions about exploiting human effort behind these advanced technologies. From a technological determinism perspective (McLuhan 1964), which suggests that technology drives societal change, the critical issue is how various human interests and investments shape this emerging technology.

Understanding Generative AI

Artificial intelligence is the simulation and augmentation of human intelligence processes by machines, especially computer systems, which includes learning, reasoning, and behavior (see the chapter "Transformation").

Generative AI, however, extends these abilities by actively creating original outputs—text, images, music, voice, video, design, code or model—derived from vast datasets, augmenting traditional AI's predictive and optimization capabilities. Unlike conventional AI, which focuses on interpreting and managing existing information, generative AI builds complex and dynamic future AI ground by generating synthetic outputs, with a significant impact across several industries. This shift expands AI's reach into creativity, innovation, and hyperpersonalized experiences. The generative AI is detectable if techniques or systems can identify outputs that AI has generated, distinguishing them from human-created content. However, this distinction is becoming increasingly challenging due to the intensifying growth of human-machine interactions and collaborations.

Generative AI represents a network of ideas and applications pivotal to technological evolution. It involves learning from data to create artifacts resembling the original inputs (Gartner Futures Lab 2023), linking to creativity. This technology stands at the intersection of the human and the machine, playing the roles of analyst, creator, and an unseen force within expanding systems. However, GenAI is a *disruptive technology* that significantly alters how industries and markets operate. With the ability to displace several established practices, GenAI prompts, even demands, a reevaluation of policy and operational frameworks.

Generative AI employs cutting-edge AI technologies to autonomously generate synthetic content, transforming fields like media, design, and data synthesis (Feher 2024a, Kruger 2023, Towards AI Editorial Team 2021). It also remixes these into redesigned formats. Currently, in the first phase of GenAI, notable uses include converting documents to slides, auto-layout adjustments, automated email replies, grammar checks, thread summaries, and enhancing videos and voices in real time (Nouri 2023, Monge 2023). These popular services streamline the creative process and democratize access to innovative tools (Pragmatic Coders 2023).

Certain services have started to provide specialized models for logos, illustrations, photorealism, and user-driven AI model training (Tollerer 2023). These kinds of combinations force more complex creative applications of GenAI services. Additionally, GenAI models can generate new output from random noise, converting them into synthetic training data for artistic, musical, and textual creation (Rouse 2023). The burgeoning startup scene in this market positions GenAI as an essential tool for expanding surveillance systems. Given GenAI's impact across various sectors, from pharmaceuticals to e-commerce (Fui-Hoon Nah et al. 2023), this technology serves as a broad interface for AI services across societal and economic functions, paving the way for a future of AI-to-AI systems learning from blended real-world and synthetic solutions.

Experts generally concur that generative AI holds transformative potential across industries, reshaping the conception and production of content, particularly in fields reliant on creativity, innovation, and knowledge processing (Feuerriegel et al. 2024). Yet, some view GenAI as a mere tool or "toy" (Bogost 2022) and not as a groundbreaking shift that redefines creativity and production boundaries. Considering users' varying access, objectives, experiences, and perspectives, these views can be acknowledged in parallel if GenAI permeates various fields, including advertising, journalism, film, music composition, and content creation, challenging traditional creative methods in different ways. GenAI permeates various fields, including advertising, journalism, film, music composition, and content creation, challenging traditional creative methods and offering unparalleled efficiency and innovation. Gregorio (2023) argues that GenAI is a disruptive force in this way, redefining creativity and content production. This view challenges conventional notions of authorship and creativity, positioning GenAI as a collaborative partner in the creative process. Nevertheless, the drawbacks of such a "revolutionary" narrative should be weighed, advocating for a balanced or critical perspective, especially in light of the risks and uncertainties.

Moreover, GenAI is a technological framework that evolves with human knowledge production, information creation, adaptability, and interaction with historical and current data sources. As Hodgkinson and Jackson (2023) explain, GenAI's technological foundation is built on sophisticated machine learning models and extensive data analysis. It utilizes deep learning and transformers not just to comprehend data but to reimagine it, producing original outputs reflective of learned data patterns. This approach is more neutral and aids critical thinking more effectively than merely focusing on specific creativity issues or how GenAI outputs are judged and interpreted across different sectors and applications.

The core principles of data-driven creativity and adaptability are crucial in defining generative artificial intelligence. GenAI goes beyond simple data processing algorithms; it embodies a dynamic cognitive process characterized by learning, reasoning, and production, steered by discernible data patterns. This cognitive process uses its knowledge base to produce varied outputs driven by human needs. GenAI, especially in models like Generative Adversarial Networks (GANs) and Variational Autoencoders (VAEs), which are trained on vast datasets, moves beyond data understanding to replication and reinterpretation, often surpassing human capabilities in speed, thought, and knowledge synthesis.

The critical aspect of GenAI is its learning and adaptability. Exposure to more data improves the realism and sophistication of its outputs. This continuous learning involves refining the AI's internal mechanisms, bridging the gap between its creations and real-world examples. GenAI's adaptability is evident in diverse applications, from image and music generation to drug discovery. For example, in films, games, and other creative industries, foundational models have been used to enhance computer graphics and streamline content creation. These large-scale, pre-trained AI systems serve as a base for fine-tuning across various tasks. Leveraging massive datasets and general representations, they showcase the versatility of generative AI. However, GenAI's ascent is accompanied by several challenges in its embeddedness in social-economic structures, necessitating a thoughtful exploration of this technology's responsibility, regulatory, ethical, or policy dimensions.

Considering all these critical narratives, GenAI's societal challenges are suggested to be approached pragmatically in the context of technology-driven creativity or adaptability. At a Harvard Business School Alumni conference, Ethan Mollick introduced the concept of *co-intelligence*, demonstrating generative AI's real-time collaboration with human agents handling complex tasks. It involves a collaborative relationship between humans and AI, where AI assists in tasks like content creation or decision-making, automating complex processes to augment human capabilities (2024). This pragmatic approach illustrates co-intelligence, positioning GenAI as a collaborative partner, like a co-pilot shaping the future of society, with the capacity to efficiently handle increased workloads and complex tasks without compromising performance or quality.

Mechanics and Models

Generative artificial intelligence operates through a sophisticated blend of models and processes, changing how machines comprehend, interpret, and generate content across diverse mediums. Hodgkinson and Jackson (2023) emphasize the variety of AI models foundational to GenAI, spanning text, image, code, video, and 3D model generation. These models, trained on extensive datasets, utilize advanced machine learning techniques like transformers to learn and replicate complex patterns. Lockett (2023) and Hodgkinson and Jackson (2023) discuss the training of these GenAI models, stressing the crucial role of large datasets in enabling AI to learn patterns, grammar, and semantics. This is evident in models like GPT, trained on vast quantities of text data to produce coherent, contextually accurate responses.

The core architecture of generative AI models includes several vital components. GANs, composed of a generator and a discriminator, excel in creating data and content that resembles real-world instances through an adversarial process. AR-CNNs enhance network performance, while Variational Autoencoders (VAEs) are essential for encoding and decoding data, useful in applications like image generation and text-to-speech conversion. Transformer models, especially GPT, use attention mechanisms to generate contextually relevant text, profoundly influencing text generation. Recurrent Neural Networks (RNNs) and a variant of them, Long Short-Term Memory (LSTM) networks, adeptly handle sequential data like text and music, excelling in language modeling and music composition due to their long-sequence memory capabilities.

These models hold significant potential for creative industries but come with risks and uncertainties. For instance, *AI drift*, a phenomenon where AI systems gradually deviate from their original goals due to changes in data, environment, or interaction, may occur with new training data or updates. This leads to decreased output quality or shifts in model objectives (The AI Whisperer 2023). Thus, significant changes in real-time GenAI systems can render services more vulnerable, less accurate, or less trustworthy.

Smith (2023) suggests employing humor as an unconventional test for complex models. Humor can evaluate the capacity to comprehend and replicate human intricacies. With this method, humor poses a challenge in understanding human expectations and producing unexpected, witty content critical in creative industries and media. While this approach is not yet widespread, it represents one of the alternatives to developing more sophisticated GenAI in the near future. Likewise, reinforcement learning can also enhance GenAI performance (Gregorio 2023, The AI Whisperer 2023). In this paradigm, models learn decision-making by performing actions to achieve goals and receiving reward feedback. However, Reinforcement Learning from Human Feedback (RLHF) confronts the fragility of human evaluation as AI systems grow more complex (The AI Whisperer 2023). The difficulty in accurate assessment can compromise RLHF's effectiveness, contributing to AI drift. A "scalable oversight" is proposed to mitigate this, where smaller, specialized AI models oversee and regulate other AI systems. This ensures that they remain manageable and human-trainable through RLHF.

These models and mechanisms represent multimillion-dollar technological ventures (Gregorio 2023)—not only mechanics to refine decision-making or scale complex tasks. Considering this business potential, these advancements in GenAI signal a profound transformation in content creation and dissemination, with far-reaching impacts on several industries and societal structures in the short term. However, the generative transformation requires regular maintenance for this potential. The concept of degenerated generative AI is emerging to highlight how these systems deteriorate in performance and output quality over time due to outdated data, reinforcing biases, or lack of updates—even if the real-time data is also part of the input. This regular maintenance should thus be part of the generative AI business models to keep its potential.

From Turing Test to Turing Trap

The Turing Test, historically a benchmark for AI's human-like intelligence, proposes an imitation game to evaluate AI capabilities. Its relevance, however, is increasingly questioned in the current era of sophisticated generative AI and the ubiquitous questions on platforms, such as "Are you a robot?"

Devised by Alan Turing about 75 years ago, this test aimed to discern if a machine could emulate intelligent behavior indistinguishable from that of a human. It was a fundamental criterion for evaluating digital and AI advancements for decades. However, as Mir (2023) pointed out, the recent evolution of AI models like ChatGPT has somewhat diminished the Turing Test's applicability in contemporary AI. According to several narratives, these models' capacity to mimic human conversation has outmoded the Turing Test. Turing originally envisaged a less than 70% accuracy rate for his imitation game by 2000, a benchmark now surpassed by advanced computing capabilities. The recent AI assessments have expanded beyond text-based interaction to include text-to-image, text-to-video, and various sensing and recognition systems. This includes several specific explorations; for instance, human-AI co-creation of video podcasts and their generated teasers (Wang et al. 2024) or image-based, multi-pose, and image-to-video techniques for marketing materials (Islam et al. 2024) to reach broad audiences.

Beyond Mir's perspective, further reevaluation of the Turing Test is also necessary. As Browning and Lecun (2023) observe, AI's limitations in grasping concrete information often stem from language's inherent abstractness and ambiguity, not just technological limitations. Over recent decades, economic and policy shifts driven by globalization and digital transformation have expanded AI applications beyond imitating human behavior to include sensory perceptions and visual recognition. These are crucial for diverse goals like business efficiency and combating fake media.

In the Turing Test reevaluation context, Erik Brynjolfsson, a Stanford Human-Centered AI professor, has presented the most popular critical viewpoint known as the Turing Trap (Brynjolfsson 2022). He underscores the impact of information technology on business strategy, productivity, and performance, suggesting that merely aiming for human-like AI intelligence might be insufficiently ambitious. Brynjolfsson posits that AI should strive to surpass human cognitive abilities in specific areas, akin to how calculators outperform human mathematical capabilities. In other words, he advocates for AI to augment, rather than replace, human capabilities, to expand the creation of new goods and services. This approach correlates with widespread AI technology narratives of human augmentation, forcing expectations beyond the Turing Test.

Incorporating both the Turing Test and the concept of the Turing Trap offers a more nuanced evaluation of generative AI. The Turing Test advances AI development by assessing its ability to emulate human behavior. In contrast, the trap-based approach highlights the concept of augmentation for the human-AI ecosystem. This dual perspective facilitates a balanced assessment of AI's societal and economic impacts. However, the Turing Test's historical significance remains notable. Its original goal of pushing machine capabilities still echoes AI's ongoing evolution. Currently, the narrative is shifting from mere imitation to innovative augmentation, urging us to envisage AI's social and emotional intelligence development.

Stochastic Parrots

The hype of generative AI faces the concept of AI systems being mere *stochastic parrots*. This term was introduced by Emily M. Bender, Timnit Gebru, Angelina McMillan-Major, and Margaret Mitchell in their influential paper "On the Dangers of Stochastic Parrots" (Bender et al. 2021). This paper became a central factor in the controversy leading to Gebru's dismissal from Google, where she was a key figure in the ethical AI research team (Hao 2020). The stochastic parrot concept encapsulates a significant dilemma in AI development: the tendency of AI models to mimic or reproduce content without genuinely understanding its underlying meaning (Bender et al. 2021). This metaphor for AI's limitations has become viral in the social sciences and humanities since 2023, when generative AI concepts emerged.

Thus, the phrase *stochastic parrots* currently underscores a core limitation in GenAI's comprehension and creativity. While generative AI models, including ChatGPT, can produce convincing and often impressively coherent content, their understanding remains model-driven, not intelligence-driven. Sanjeev (2023b) observes that these models often need a deeper comprehension of complex concepts or contexts, resulting in outputs that may be grammatically sound but contextually inadequate or nonsensical. Furthermore, the generation of information without genuine understanding and the production of expected responses can lead to confabulation and hallucinations (Gujral 2023, Sinatra & Hofer 2023). This issue represents a broader challenge: emulating human intelligence without achieving true cognitive depth.

There is also criticism of the *authoritative style of GenAI outputs*, defined by the confident and often definitive nature of AI-generated responses, contrasting with human experts who acknowledge the limits of their knowledge and embrace uncertainty (Kruger 2023). This presents a significant communication and information consumption challenge, particularly when recognizing uncertainties or limitations. The suggestion in this case is to encourage GenAI systems to admit ignorance or lack of relevant data in specific ethical scenarios.

Regarding GenAI technical detection, one suggestion is *feature engineering*, which is the process of selecting, modifying, or creating new features from raw data to improve the performance of machine learning models. A further one is *watermarking*, a technique used to embed a hidden marker or logo in a document or data to protect copyright, verify authenticity, or signal ownership. A more extreme detection method is *poisoning training data*, intentionally altering it to compromise and disrupt future content-generating AI models. However, hidden markers or poisoned data are less favored due to the difficulty of detection and the risk of malicious misuse (Heikkila 2023). Societal and business trust, combined with a culture of skepticism and independent verification, can more effectively shape understanding of the probabilistic nature of GenAI responses. This approach encourages users to evaluate AI outputs critically and fosters a more informed and cautious interaction with GenAI.

The news industry, in particular, faces significant challenges due to GenAI. Mike Ananny, an associate professor at USC Annenberg, and Jake Karr, deputy director of New York University's Technology Law and Policy Clinic (2023), argue that GenAI undermines journalistic control over language and content. The news industry's historical adaptation to technological changes now faces a new test, leading to increased dependence on tech corporations. This dependence raises concerns about compromised journalistic integrity, as it may result in content that does not adhere to journalistic standards of truth and public interest. While there are concerns about how GenAI could undermine key journalistic values, it's also worth considering how it might enhance journalism through automated tasks, AI-supported data analysis, and preliminary report drafting, allowing journalists to focus on more critical aspects of their work.

The current state of AI, characterized by feeding data and prompts into machines, is a temporary phase in AI's evolution (Heaven 2023a). Users are effectively teaching these synthetic systems, which lack biological limits. Future developments are expected to move beyond this basic input-output model, potentially evolving toward a deeper understanding and processing of information. Advancements in prompt engineering signify this transition and indicate an ongoing evolution in GenAI (as detailed in the chapter "Transformation").

The concept of AI as stochastic parrots extends beyond technical shortcomings to include ethical and societal implications. Initially, biases in training data are replicated in the first stage of GenAI. However, a concerted effort to reduce harm and prejudice is leading to a shift toward a more homogenous synthetic culture in its second stage. The risk of misuse in creating convincing mis/disinformation or deepfakes raises questions about responsible use of technology. Subsequently, GenAI will undergo fine-tuning through tech-human reviews to mitigate mis/disinformation. This process is critical, especially if the news industry, which plays a pivotal role in societal and economic balance, is to maintain public trust.

In line with this, GenAI also significantly impacts intellectual property and creativity. As AI generates content, it questions the boundaries of "prosumer" roles in human-GenAI interactions. Everything created, from prompts to texts, images, videos, games, music, news, or code, instantly becomes part of GenAI. It transforms into outputs for other users or even interacts with the prompter—as a non-repeatable interactive process. This constant exchange blurs the lines between authorship, intellectual property rights, and roles in creation and transformation.

Looking forward, a stochastic parrot causes controversial human experiences but demonstrates a catalyst for transformative change. This signals the need for advanced AI models capable of deeper understanding, reasoning, and ethical responsibility.

Societal and Business Impacts

Semantic, context-based generative search and content have profound societal impacts, driving widespread changes in the functioning of information generation and content-driven websites (Gregorio 2023, Kantrowitz 2023, Kennedy 2023, Nouri 2023). GenAI, integrating human interaction with media-communication-information structures (Feher et al. 2024), is transforming our engagement with generated content and redefining how we communicate, create, and consume information.

As Mir (2023) has discussed, the diminishing relevance of the Turing Test signals a significant shift in AI's societal role. The human-like quality is no longer the gold standard for AI technology, particularly considering the risks of biases or errors due to less diverse training data. These issues raise societal-ethical concerns, including problems with data labeling and exploiting low-wage workers (Lockett 2023, Heaven 2023b). However, tailored data labeling can foster a society-fit and human-centered enhancement for GenAI, addressing inequality and harmful content. For instance, GPT-4 refuses to respond to prompts that propagate harmful stereotypes (Chen et al. 2023), marking a significant shift from the 2016 incident in which Microsoft's AI chatbot Tay was terminated for spreading hate speech. This evolution highlights a move toward responsible AI development and deployment, which is crucial for building public trust in technology.

Yet, societal trends indicate a growing *techno-narcissism*, a critique of the excessive fascination with and overvaluation of technology, often at the expense of human values and social relationships (Galloway 2023). This has led to an increased dependency on GenAI technology and overestimating its capabilities, often overlooking its limitations and ethical ramifications. For example, replacing human journalists with robot reporters for cost-efficiency overlooks the nuances of ethical and critical thinking inherent in human reporting. Such radical changes impact audience expectations and trust in unpredictable ways. The OECD (2023) also highlights a significant language bias in GenAI, dominated by English, raising concerns about preserving diverse linguistic and cultural identities.

Despite these challenges, GenAI is transforming the internet and information consumption. The key question is the scale of this change, and we need a clearer understanding of how the influx of generated low-quality web content will impact search traffic. Therefore, deeper exploration is needed into generative AI's scalability—its ability to manage growing complexities efficiently—and its *scaling* potential across industries. GenAI-driven content ultimately transforms cultural, ethical, and economic systems, reaching a *societal scale of generative AI*. Learning about these changes has triggered two leading narratives in recent times. Leading business conferences are advancing a narrative of radical transformation and scalability in content and media, paralleling the enthusiasm for sustainability from AI-generated green marketing campaigns to eco-friendly content strategies. Others, primarily academic narratives, still underline the concept of the platform economy (Kenney & Zysman 2016), indicating potential stability for established business models in data and content markets. This stability is contingent upon leading tech companies retaining their competitive edge through the accumulation of vast resources, data monopolization, and significant market power, tactics that have enabled them to maintain market dominance for decades and set industry standards, often limiting the entry of new competitors. However, this presumed stability is subject to ongoing pressures from technological innovations, shifts in user behavior, and evolving regulatory environments, where new market entrants, especially in generative AI, have the potential to disrupt and surprise the market of digital transformation, challenging these dominant positions in the long term. An example of the latter is the rise in licensing deals, reminiscent of the early days of book publishing when authors and publishers negotiated rights for content usage and distribution. Generative AI companies offer substantial sums to the news and content industries for using their data in model training (David 2024) and providing tools for efficient content production and filtering of fake content. These early legal actions and restricted access to public content due to intellectual property concerns further catalyze innovative business models' emergence.

These societal and business impacts raise questions about the future of job roles in cultural-creative fields. As Kruger (2023) discusses, generative technology poses concerns, especially in areas traditionally dominated by human creativity, such as writing, art, filmmaking, advertising, and design. While GenAI opens new avenues for creativity, it challenges existing employment structures and skill relevance. Meanwhile, creative works involving GenAI are spurring innovative practices, such as artists revealing GenAI's role in their work or grappling with communicating the complexities of their AI-assisted creations.

In conclusion, GenAI is upending media and information-communication, transforming content creation and dissemination, and introducing responsibility issues. It risks fostering techno-narcissism and deepening cultural erasure, threatening jobs in creative fields with automation. The market also sees tech giants' dominance challenged, spurring licensing deals that commodify data, yet this raises uncomfortable questions about control, innovation, and whether society is ready to confront the trade-offs of GenAI's unchecked rise.

Entertainment and NewsTech

AI tools like ChatGPT are transforming content creation; citing an example, 77.92% of participants in Ibero-America utilize generative AI for tasks within the news industry (Apablaza-Campos et al. 2024). While this integration enhances efficiency and allows media professionals to focus more on in-depth investigative work, it also brings ethical concerns about the credibility and bias inherent in AI-generated content (Vicsek et al. 2024). There is a risk that AI-generated news could exacerbate the *commercialization and refeudalization* of public discourse, echoing Grosswiler's (2001) warnings about technology reducing complex discussions to mere entertainment. Moreover, the rise of algorithm-driven augmented journalism, or *NewsTech*, illustrates a shift from traditional news production and dissemination to data-driven or algorithm-driven personalization (Feher 2024a). This reliance on AI in content creation might lead to a more engagement-focused media ecosystem, potentially at the expense of critical analysis.

In the entertainment industry, including music, film, and gaming, generative AI introduces both innovative potential and significant pitfalls. It enables new forms of creative expression, like AI-generated impacts and immersive virtual experiences, but also raises concerns about originality, authorship, and the potential for misuse. For example, AI-based affective music generation (AI-AMG) offers personalized, real-time emotional adaptation, greatly transforming user experiences in gaming and interactive media (Dash & Agres 2024). This advancement sparks a critical discussion about the authenticity of AI-generated music, as its ability to closely replicate human creativity challenges traditional ideas of originality and authorship, potentially shifting our understanding of what constitutes genuine artistic expression in GenAI times. Another example of generative AI in the film industry is automating tasks such as scriptwriting and visual effects, which can streamline production processes, reduce costs, and challenge the traditional roles and values of human creators (Bender 2024). This shift could diminish the demand for creative professionals in the future artistic-cultural labor market.

The intersection between NewsTech and entertainment in generative AI includes shared challenges such as intellectual property rights, datafication, and bias. In NewsTech, AI-generated content forces professionals to balance the thin line between enhanced content and the risk of reinforcing ideological biases (Breazu & Katsos 2024). For the entertainment sector, the concerns lie in the potential erosion of human artistic input and the regulative and ethical implications of using AI to create new works. Both fields confront how generative AI redefines authorship and representation, suggesting the necessity for a comprehensive framework to address these complexities.

From a critical perspective, NewsTech can improve fact-checking and transparency (Musi et al. 2024); they also pose risks of ideological biases and may reduce nuanced reporting, particularly concerning marginalized groups. In the entertainment industry, generative AI offers endless possibilities for immersive storytelling and personalized content creation; it also produces homogenization, as algorithm-driven production may prioritize commercial appeal over diverse, culturally rich narratives, potentially sidelining marginalized voices and limiting creative risk-taking. Therefore, several AI guidelines stress the importance of human oversight to prevent errors and biases in AI-generated content. However, the predominance of the Global North in formulating these guidelines raises concerns about power asymmetry (de-Lima-Santos et al. 2024). This underscores the need for an inclusive and globally-minded approach to generative AI.

Both NewsTech and entertainment also face the challenge of AI-generated mis/disinformation, which has the potential to significantly shape public opinion (Ananny 2024a, 2024b). This underscores the need for a normative shift in AI literacy, advocating not just for responsible but radically creative uses of the technology. The power to transform media in the public interest lies in making bold, informed choices about GenAI's role—ensuring it fosters diverse, critical engagement and challenges the dominance of profit-driven, homogenized content (Deuze & Beckett 2022). Therefore, the *creator economy, especially* in NewsTech and entertainment, refers to a transformation where individuals use AI tools to generate content, products, and services, fostering new avenues for creativity and monetization while introducing both unique challenges and opportunities.

Future of Generative AI

Generative AI is poised to break creative barriers, forcing a radical redefinition of art, authorship, and responsibility—especially in light of OpenAI's original mission to prioritize humanity's welfare over profits (even if today it already predominantly serves profit-driven goals). Yet, as Biggins (2017) summarized, ethical considerations, often overshadowed by business interests, have struggled to keep pace with rapid technological advancements. This business-driven technological shift generates significant macro and micro societal challenges. It affects democratic processes, labor markets, environmental sustainability, and the everyday life of communities and individuals. It also changes professional roles and knowledge production and potentially leads to cognitive atrophy due to decreased human neural connectivity, as highlighted by Sætra (2023).

The expansion of generative AI into numerous sectors highlights its potential and societal challenges (Dwivedi et al. 2023). As AI capabilities expand, striking a balance between technological growth and human values becomes increasingly vital. This equilibrium is key to ensuring that AI positively impacts society (Fui-Hoon Nah et al. 2023). Thus, technological advancements in GenAI are expected to bring enhanced multimodal capabilities and improved natural language understanding, potentially merging with robotics and advanced learning algorithms, as forecasted by Google DeepMind (2023).

The *future of generative AI* promises an evolution in how we create or conceptualize something new, produce knowledge or services, and interact with content, blending synthetic and human-centered approaches. This perspective embraces a holistic view of human existence concerning advanced digital technologies and a human-AI ecosystem, fundamentally challenging the essence of human originality as AI begins to simulate emotions, ideologies, and cultural narratives. While the echo chamber effect underscores the related limitations (see chapter "Transformation"), it also enables the amplification of conspiracy theories. As the amount of generated synthetic content increases radically, the probability of these extreme theories gaining traction within echo chambers rises, which can contribute to radicalization and deeper societal divisions. This dynamic illustrates the need for ongoing scrutiny of how generative AI shapes public discourse and influences belief systems.

One of our studies balanced these foresights, exploring a normative scenario of AI futures through participatory research with leading experts (Koves et al. 2024). The research findings introduced *entangled AI*, a concept drawn from quantum physics to highlight the deep interconnectedness between AI, humans, society, and nature. This concept emphasizes a balance of social well-being, justice, and environmental resilience. This framework envisions AI as an integral part of a global collective consciousness intricately linked with various aspects of human existence.

Applying the concept of entangled AI, GenAI allows us to draw inspiration from a complex interconnectedness. This foresight pushes human-GenAI co-intelligence to evolve a collaborative partnership that interacts with social networks, natural environments, and reflective processes. The benefit of this approach is that it enables GenAI to enhance human experience by integrating diverse, nuanced inputs. However, it also risks blurring the lines between human autonomy and AI influence, potentially diminishing independent thought and agency, probably in the creative control of tech conglomerates.

Impact Project

When the FBI deployed its targeted AI system to confront a GenAI-driven fraud scheme, it marked a symbolic moment in the battle between emerging technologies. During the investigation, agents discovered that a musician had used generative AI to produce thousands of fraudulent songs while employing automated bots to artificially inflate streaming numbers, ultimately generating over $10 million in illegal royalties (Noyes 2024). This case revealed significant weaknesses in music streaming platforms, where AI-generated content blurred the lines between authenticity and ownership, posing severe challenges to intellectual property and media integrity.

Throughout the investigation, the FBI utilized advanced Natural Language Processing (NLP) and pattern-recognition algorithms to monitor streaming data in real time, identifying anomalies and uncovering extensive bot networks. This sophisticated approach demonstrates the increasing complexity of AI-driven crimes and the necessity for law enforcement to employ at least as much advanced technology in response. In a broader sense, the case highlights the growing AI-to-AI interplay, in which technology acts both as a tool for creation and a mechanism for fraud detection. It showcases how law enforcement must adapt to rapidly advancing technological threats across a variety of industries by spotting patterns and anomalies in large datasets and enabling more accurate and efficient prediction and prevention of fraud.

Federal authorities described the case as "brazen fraud." It is considered the first to have involved artificially inflated music streaming, setting a significant precedent for the law enforcement approach to AI-driven crimes. This case provides valuable insights into how the misuse of GenAI challenges AI governance. Law enforcement can develop systems that preemptively identify and reduce fraudulent activities by continuously refining AI detection algorithms and integrating them into monitoring mechanisms. In the context of streaming fraud, for example, future AI systems could distinguish between human-driven and bot-driven behavior, learning from past fraudulent cases to stay ahead of increasingly sophisticated schemes.

AI-versus-AI systems are emerging as the next frontier in addressing the growing complexity of GenAI-driven harm. The case highlights the demand for advanced AI systems to detect and mitigate fraud and misuse in generative and synthetic content. This dynamic—AI detecting AI misuse—represents a new phase in governance, where real-time adaptability and feedback loops are critical. The precedent set by the FBI underscores the effectiveness of adaptive governance frameworks that allow AI systems to learn from fraudulent misuse and continuously improve, ensuring that generative AI remains a tool for positive innovation rather than a vehicle for harm.

Reflective Art

Generative AI transcends its role as a mere tool for creating art; it also acts as a mirror to society and culture. Examining projects that employ GANs reveals a dynamic interplay between technology and art. These examples demonstrate how generative AI reflects our collective intellect and cultural norms and serves as a creative force, forging new artistic domains that extend beyond the limits of conventional creativity. Investigating three significant examples, we see their enduring impact and societal relevance.

Firstly, the *Mirror Test and AI Reflection* concept, discussed by Vincent (2023), offers an insightful parallel. This concept compares AI's capacity to mirror human knowledge and language with the mirror test from behavioral psychology, which evaluates animal self-awareness. This analogy is crucial for comprehending GenAI's role. AI doesn't merely duplicate information; it mirrors our collective intelligence and linguistic nuances, prompting deep contemplation about its potential self-awareness and how we perceive this technology. This reflection is a persistent marker of AI's evolving capabilities and our continuously transforming relationship.

Secondly, Refik Anadol's *Unsupervised*, exhibited at the Museum of Modern Art (MoMA), was highlighted by Diehl (2023). This synthetic artwork employs AI algorithms to convert images and metadata from MoMA's archives into abstract, fluid visuals. This artwork breaks free from traditional artistic conventions, crafting a novel domain of visually arresting experiences. Its significance lies not just in its technological innovation but in how it redefines audience engagement with art, marking a milestone in the seamless integration of GenAI into artistic expression.

Lastly, the application of Generative Adversarial Networks (GANs) in Beko's *Sustainability Film* (Beko 2022), as described in their behind-the-scenes release, represents a momentary yet significant use of generative AI in creative media. This project aimed to promote sustainability without showcasing products, using AI-generated nature images to create a sense of connection and motivate viewers toward sustainability. Beko combined AI-generated visuals with live-action footage for their promotional film. This blend exemplifies AI's potential to enhance and transform creative media production. While the film is ephemeral, its approach is a harbinger for future projects in the industry, underscoring AI's growing significance and versatility in various creative endeavors.

These instances illustrate GenAI's capacity to mirror and augment human creativity, potentially redefining our understanding of art, media, and intelligence as contemporary snapshots of GenAI in art. As Epstein et al. (2023) assert, the unique aesthetic qualities of these works are indisputable, even though issues like authorship and future job impacts remain unexplored. Institutions like MoMA or the Venice Biennale have recognized the aesthetic value of GenAI artwork, acknowledging its place in the annals of art history.

However, these artworks also represent transient reflections of the current state of generative AI, each marking a milestone in the evolution of this technology within the art world. They underscore AI's dual function as a mirror to our cultural and intellectual milieu and as a sculptor of novel, unexplored artistic frontiers. As Lev Manovich (2023) noted, we are witnessing a significant shift in media creation, with culturally significant artifacts transforming the nature of creativity.

Conclusion

As generative AI shifts us into an era of synthetic complexity, we are confronted by its transformative potential and profound ethical stakes. No longer limited to static, rule-based interactions like elementary chatbots, generative AI redefines the foundation of human-machine relationships. By challenging conventional models such as the Turing Test and confronting the Turing Trap, adaptive agents, real-time hyperpersonalization, and creative human-machine content generation drive a dynamic shift. This evolution of *static vs. dynamic human-machine interaction* positions generative AI not merely as a tool but as an evolving co-intelligence, reshaping our understanding of intelligence, agency, and interaction.

The notion of generative AI as a "sociopath GenAI," a term highlighting its potential to mimic without proper comprehension, contrasts sharply with its role as a catalyst in the AI-driven creator economy. This duality speaks to the transformative power of generative AI, which has reshaped human abilities and emerged as a crucible of innovation. Yet, a disruptive technology such as this is not without its pitfalls. Issues like AI drift, data bias, and the need for inclusive and innovative approaches underscore the complexity of this evolving field.

Considering the societal and business impacts of generative AI, it is evident that this technology is not just a tool but an authoritative style shaping several industries and artworks. As GenAI advances, it challenges traditional notions of creativity, labor, and human contribution by automating tasks and generating new forms of content, speeding up information industries and contextualization. While it offers opportunities for innovation and efficiency, it also raises concerns about the erosion of human agency, authorship, and the value of work. Although many narratives suggest a balanced approach between these forces, this perspective often reflects a desire to preserve existing societal structures and power dynamics rather than truly considering the profoundly transformative potential of GenAI for the next generations of innovation.

Scholarly Debates

For Cultural and Social Sciences

- How does generative AI transform traditional concepts of authorship and creativity in humanities and social sciences?
- What cultural narratives are emerging regarding generative AI, and how do they influence public perception or societal expectations?
- How does generative AI influence the portrayal of diverse cultures and their perception in media and art?
- How do concepts like the Turing Test, sociopath GenAI, and techno-narcissism shape philosophical debates?

For Science, Technology, and Engineering

- How do generative AI's mechanics and models differ from traditional AI, and what are the resulting implications?
- How does generative AI's potential for "AI drift" impact the long-term reliability and safety of AI systems?

- How can watermarking and detection techniques help reduce the risks of generative AI creating harmful or deceptive content?
- How does generative AI challenge the principles of the Turing Test, and what does this mean for GenAI's future development?

For Business, Policy, and Innovation

- What are the challenges and opportunities of integrating generative AI into corporate innovation strategies?
- How does generative AI influence consumer behavior, and what are the implications for marketing and advertising strategies?
- How can policymaking balance innovation and standards to ensure the responsible but competitive development of emerging generative AI technologies?
- How can policymakers address the potential for generative AI to disrupt existing industries?

Strategic Takeaways

For Academia

- Curriculum Integration of Generative AI: Incorporate comprehensive courses on generative AI, focusing on its mechanics, models, competitiveness and responsibility considerations, and societal impacts to prepare students for a rapidly changing GenAI context.
- Research on Turing Test and Stochastic Parrots: Focus on generative AI's stochastic parrots behavior while assessing its performance through the Turing Test for human-like intelligence.
- Interdisciplinary Collaboration: Encourage collaborative projects across computer science, ethics, psychology, and sociology to tackle complex issues of GenAI.
- Encourage Hackathons: Explore generative AI's potential to solve societal issues; enable students and researchers to develop innovative tools and apply GenAI practically and ethically.

For Industry

- Innovative AI Applications: Explore generative AI applications in content creation, marketing, and product development; consider their potential and limitations in a disruptive context.
- GenAI Deployment and Training: Align GenAI usage policies with organizational values and introduce upskilling training on its opportunities and challenges for effective implementation.
- Strategic Planning for GenAI Integration: Develop strategies for integrating GenAI in business processes. Consider its impact on efficiency, cost, workforce, and GenAI-driven creator economy.

- Understand GenAI's Industrial Frameworks: Focus on automation, content generation, data analysis, and personalized solutions for efficiency and innovation.

For Policymaking

- Regulatory Frameworks for Generative AI: Formulate regulations for generative AI development and usage, focusing on ethical use and data privacy.
- Public Education on GenAI: Initiate public education programs to inform citizens about GenAI's benefits, risks, societal role, and critical narratives.
- Fostering Responsible GenAI Innovation: Promote responsible GenAI innovation by balancing technological advancement with societal and ethical considerations.
- International Collaboration on GenAI Standards: Encourage collaborations to develop joint standards and best practices for generative AI. Address challenges such as reliability, originality, or integrity.

3

AI MEDIA

Defining AI Media

AI media integrates artificial intelligence with content creation, distribution, moderation, and personalization, effectively merging human creativity with algorithmic processes. This integrates AI-driven social media, synthetic media, and generative media, and the convergence of these various media functions. AI mediatization changes how content is produced and consumed, ranging from robot journalism to synthetic immersive experiences, while AI-generated content demonstrates the capabilities of these technologies. However, deepfakes, misinformation, disinformation, and manipulative recommendation systems constantly challenge the integrity of information, the trustworthiness of media, and the reality of synthetic outputs.

Introduction

The evolution of media from mass communication to algorithm-driven services signifies a dynamic interplay among technology, society, culture, and business in shaping media, as we discussed at an IEEE conference on cybersecurity and privacy (Feher 2020, Mansell & Steinmueller 2020). The *new-smart-AI media evolution* and the coexistence of the related technologies (Feher 2020, 2022c) facilitate diverse human-machine interactions and mark a progression from conventional digital content consumption to complex AI-driven systems. These systems include bot journalism, personalized services, and automated decision-making, reflecting a shift toward a sophisticated, interconnected media environment where algorithmic information flow and data-driven decision-making are increasingly prominent.

According to our systematic review of socio-cultural AI (SCAI) (Feher & Katona 2021), AI media topics have been primarily focused on the past decade

DOI: 10.4324/9781003591023-4

and come from the social sciences and humanities. Studies from 607 highly-ranked journal papers from seven academic databases show a marked interest in AI-driven media beginning in 2015, with a sharp surge from 2018, paralleling other critical topics like AI in healthcare and governance. Key research areas include social media, fake news, automated journalism, the AI-driven news industry, and deepfakes, with news generation as a central theme. The prominence of these research areas underscores the pervasive influence of AI-powered media in society and culture (Winiarska-Brodowska & Feher 2025, Feher & Katona 2021). This systematically studied literature represents how media and AI converge, resulting in a significant shift in content creation, automated media production, hyperpersonalization, and distribution, impacting sectors from news to entertainment. The implications of AI on employment in the media sector, its role as a marketing tool, and the tendency to underplay AI's failures are crucial concerns. Accordingly, Shah (2024) emphasizes the need for a balanced approach to leveraging AI's capabilities, grounded in academic research and analysis.

This chapter emphasizes the pivotal role of content and news creation through computational methods and automation. However, the human element remains essential, particularly in how experts shape and envision AI's future. Despite challenges related to deepfakes and AI media ethics, a dangerous sense of high expectations and optimism remains. This chapter summarizes AI's current impact on media while also interpreting the optimism surrounding the human factor in shaping its future.

From Metamedia to AI Media

Media is a powerful catalyst, revealing intersections between technology, social relations, economic models, political interests, and aesthetic expressions. The concept resonates with USC's "Media as SocioTechnical Systems" perspective (2024), framing media systems as active shapers of perception across technological contexts. This approach expands our understanding of AI's societal implications beyond isolated media fields, encouraging interdisciplinary insights. As Guzman and Lewis (2024) also emphasize, an inter-industry approach could offer a holistic view of AI's growing impact on the media ecosystem.

This concept is particularly relevant in Generative AI, as these technologies transform media by automating content creation, enhancing personalization and enabling real-time interaction, fundamentally altering the dynamics between creators, audiences, and algorithms. By integrating AI-driven insights and capabilities, media can adapt individual preferences and societal trends at advanced scales, which raises critical questions about agency and the balance of human and machine influence in cultural and social narratives.

Applying similar concepts, the term metamedia has emerged over the last decades, emphasizing ubiquitous computing where various forms of media are integrated and transformed through interaction, adaptability, and narrative (Jensen 2010). This concept also involves a computing transformation of media processes and contexts (Campos-Freire et al. 2016). Unlike traditional multimedia, which combines different media formats, metamedia transcends these boundaries by creating unified platforms that foster active engagement and customization.

While metamedia overlaps with AI media, *metamedia* is not entirely synonymous with it; instead, it represents a broader framework that encompasses both AI-driven content creation and the dynamic interactions between media forms and their individual or institutional users. The transformative potential of metamedia aligns closely with the principles of metamodernism, a cultural framework that oscillates between modernist ideals and postmodern skepticism, embracing complexity and hybridity in a fragmented world (Vermeulen & Van den Akker 2010).

Metamedia goes beyond merely merging media forms, pushing past Jenkins's cultural convergence theory (2006) into a new paradigm. It signifies a transformative shift in media production, experience, and interpretation driven by AI-generated content and hyperpersonalized consumption. Yet, the rapid proliferation of AI-generated media intensifies information overload, making it increasingly difficult to distinguish credible sources from fabricated content. This surge raises profound questions about *media credibility*, emphasizing the need for accuracy, transparency, and ethical standards, particularly in generative or synthetic media production.

Moreover, Floridi (2023) argues that generative technologies, by perpetuating biases, spreading mis/disinformation, and eroding trust in media institutions, risk undermining the cultural and social fabric without careful consideration. In this context, metamedia, as a concept, extends beyond mere technical integration to emphasize the social responsibility of media creators and users, to support social cohesion. Thus, the concept of metamedia promotes active, engaged media literacy, encouraging users to interact with and shape content while raising awareness of underlying algorithms and potential biases. To mitigate these risks, the metamedia framework must include more robust safeguards to ensure that AI media serves as a tool for empowerment rather than manipulation. As Gunkel (2018) emphasizes, promoting a more innovative and responsible media environment requires a nuanced understanding of AI agency and its impact on human culture. This metamedia concept balances the potential of cultural enrichment and prevention against manipulation.

AI media is a less abstract but more pragmatic concept of contemporary media. It refers to specific news, entertainment, social, virtual, synthetic media, or "media as a service," driven by data analytics and generative technologies. All media operations, such as production, distribution, and moderation, are included in the AI media. As our studies in AI-generated media, particularly through a systematic analysis of 2,727 records from 13 international databases, revealed, the term "AI media" has emerged to describe the convergence of social, news, artificial, fake, and service-driven media (Feher 2024b). This concept identifies fundamental challenges, including trust and ethics, privacy safeguards, creative industries, and the sustainability of AI media ecosystems.

Studying the key narratives in this research, two definitions of AI media are available. First, *current AI media* refers to social-artificial, synthetic, or fake media and media-dependent services generated by technologies like generative AI and algorithm-driven content production and distribution analytics. Key socio-economic aspects include business investments, the future of work, media-driven industries, and artistic reflections. Second, *AI media future* is expected to encompass advanced media production and services powered by recognition-behavioral-transaction technologies and sophisticated analytics. Critical challenges include AI ethics, responsibility issues, trust in AI-driven media, privacy protection, and sustainability within algorithm-based ecosystems.

GenAI is deeply connected to this concept of AI media, providing the technological backbone that drives its development. Generative technologies transform media production and consumption and fuel the evolution of AI media by enabling the creation of synthetic content that blurs the lines between human-made and machine-generated media.

Proposing a definition accordingly, GenAI media refers to AI-driven content creation using machine learning models, vast datasets, and statistical techniques to generate synthetic media that mimics human outputs. GenAI media blends human and machine inputs and outputs, transforming socio-cultural conversations and public discourses, challenging authorship and authenticity (Raley & Rhee 2023), and influencing the infrastructures of meaning-making essentials (Fischer & Jarren 2024).

Convergence of Media and AI

AI has become so deeply integrated with media that this intersection has evolved beyond simple technological innovation in media industry. Mir (2023) redefines "media" as not merely extensions in our environment but as part of the mind, directly linked and integrated into digital networks, progressing toward "neuralinked" direct mind connections. This creates networked or synthetic environments, profoundly influencing media perception and interaction. Wirtschafter and Dews (2023) further explore this convergence, examining how information traverses digital media. Their analysis of thousands of podcasts reveals significant data transformation and crowdsourced content moderation, quantifying the media's evolving scale. Their findings suggest AI's reach extends beyond social or mass media, employing advanced data science and natural language processing to analyze extensive datasets.

Networked infrastructures, crowdsourcing, and hybrid computing networks fuel and enhance AI outputs, creating more dynamic human-computer interfaces (Muhlhoff 2020). Multimodal interfaces, which integrate multiple communication methods like voice, gestures, touch, and visual elements, play a pivotal role in generative AI and AI-driven media. These interfaces make human-computer interaction more intuitive and flexible, enabling richer, more natural engagement across diverse input modalities. This improves user experience in virtual assistants, smart devices, and immersive technologies like AR/VR/XR, offering extended experiences that surpass traditional interactions.

These technologies harness human cognitive capabilities, aiding AI development and shaping digital media interactions. Complex power dynamics are embedded in this process, ranging from gamification to audience labor (Muhlhoff 2020, Fisher2010). Muhlhoff (2020) demonstrates how gamification in non-gaming contexts engages users in AI training tasks, while platforms like Google and Facebook subtly convert users into data sources. Wischnewski et al. (2022) provide another example, studying social bot interactions on X (formerly Twitter), highlighting AI's nuanced media impact.

Users show increased engagement with bots—brief *bot-engagement*—that exhibit human-like traits that align with their political views, underscoring AI's complex role in public discourse. Shaikh and Moran (2022) examine media partisanship in portraying technologies like facial recognition, revealing its influence on public perception and potential regulatory impacts. These examples indicate an increasing interconnection among social media platforms, AI systems, and digital service providers in networking users, media usage, and AI technologies.

This convergence signifies AI's integration into media ecosystems and vice versa, where media technology trains AI and AI enhances media services—as an *AI-to-AI loop*. However, this process raises challenges in compatibility and interoperability. The key questions are how AI technologies can be seamlessly integrated with existing media platforms and how AI media learning feedback can further drive convergence and challenge the media and information-communication industry.

The convergence challenge is twofold: technological and ethical-psychological. The AI4Media project, as discussed by Tsalakanidou et al. (2021), addresses these challenges, focusing on new learning paradigms, explainability, robustness, privacy, and societal impacts. Their analysis of seven industrial use cases underscores AI's disruptive role across sectors, recommending developments in automatic video summarization, audio generation, and multi-language models for trust and accountability in media.

Multifaceted psychological and ethical challenges are emerging within this convergence. Several academic studies have explored these issues, including Ali Abbasi's call for ethical governance to counter AI-amplified biases (Ali Abbasi et al. 2022), Lanier (2014) and Roberts' (2016) focus on AI's benefits over controversial content moderation, and Hao's (2021) and Sanjeev's (2023b) examination of deepfake weaponization and disinformation. Additionally, Sundar's (2020) study on how media shapes user trust in AI systems, Calice et al.'s (2023) exploration of algorithmic news curation's impact on media perceptions, and Natale's (2022) research on social behaviors projected onto AI. These all provide crucial insights for building trust in AI-driven media systems and understanding their psychological and social effects from several aspects and perspectives. The interplay between AI and media presents a significant opportunity to engage with the currently diverse and sometimes fragmented studies. It also offers a chance to converge narratives and interests within AI media research.

AI and Generative Mediatization

Historically, the concept of mediatization focused on how media technologies shape and influence society, culture, and communication. Scholars explored how digital media transformed political communication, religion, and language (Hjarvard 2013, Couldry 2008). The term reflected the dynamic interplay between media systems, social behaviors, and cultural practices. As media technologies evolved, they were seen as communication tools and active agents in shaping societal norms and personal experiences (Jansson 2015).

Deep mediatization, introduced by the German media theorist Andreas Hepp, extended this concept by emphasizing the role of data and algorithm-driven systems in transforming media infrastructures. Hepp highlighted how digital media became embedded in everyday life, influencing social and cultural processes through datafication and algorithms (Hepp 2020). This infrastructural transformation illustrated how deeply integrated media systems had become, impacting personal interactions, cultural discourse, and social relations. As AI technologies advance, however, the deep mediatization concept already falls short of capturing their real-time, dynamic effects, necessitating further conceptual development.

As more complex AI technologies emerge, *AI mediatization* further complicates the concept of mediatization, which describes the integration of AI within media processes. This evolution highlights AI's growing role in transforming traditional media ecosystems and processes. Unlike earlier phases of mediatization, AI-driven media represents a qualitative shift. In this context, AI enhances traditional media processes, disrupting and redefining them. AI's capabilities, particularly generative AI technologies, introduce a further level of automation and algorithmic complexity.

Generative AI hype has also reshaped narratives in media, challenging the traditional boundaries of authorship and authenticity (Raley & Rhee 2023) as AI-generated content blends human input with machine-driven production. AI-driven content platforms such as Alibaba's Media AI system demonstrate this shift, offering rapid, template-based content production that can meet consumer demands more efficiently (Cloud 2020), while voice assistants are emerging as critical players in news curation and delivery (Arifin & Lennerfors 2021).

Accordingly, generative mediatization refers to the dynamic interplay between AI algorithms and media systems. It describes how AI transforms traditional media production, modification, and distribution processes. This concept highlights AI's dual role as a disruptor and catalyst in media evolution (Laba 2024,

Manovich 2023). Generative mediatization builds on the deep mediatization framework by incorporating the unique qualities of generative AI, which not only automates but also personalizes media content on a large scale.

The political implications are evident in this complex AI and generative mediatization. Applying Voting Advice Applications (VAAs), where Large Language Models (LLMs) like GPT-4 are used to create dynamic political avatars, provides nuanced insights into candidates based on real-time data from platforms like X (see Mancera Andrade & Teran 2024). While these AI innovations enhance political analysis and voter guidance, they also raise concerns about bias, mis/disinformation, and the transparency of AI systems.

AI mediatization and generative mediatization significantly depart from traditional mediatization models. While deep mediatization emphasizes the quantitative and infrastructural changes brought by datafication and algorithms, AI mediatization expands this by focusing on AI's role in shaping media ecosystems. AI is no longer just a tool used within media but a fundamental actor that drives and transforms media production, distribution, and consumption. Generative mediatization complements AI mediatization by focusing on AI's transformative impact on the *content creation process*. AI's ability to generate synthetic content—for entertainment, news, or even political communication—alters the media production and challenges existing structures of authority, authenticity, and creativity.

To sum up, AI mediatization and generative mediatization are interconnected, with AI driving the technological advancements that transform media processes while generative AI transforms content production. Together, they highlight the evolving role of AI in the media ecosystem, signaling an upcoming phase in the media's relationship with society.

Robot Journalism

It signified a pivotal moment in the fusion of journalism and technology when the LA Times published an article on an earthquake written by a robot in 2014 (Schwencke 2014). This event highlighted the rise of "robot reporters," now used to cover various topics from sports to finance (Gentzkow 2020). The Guardian's collaboration with GPT-3, resulting in "A robot wrote this entire article. Are you scared yet, human?" (Guardian-GPT3 2020), further illustrated AI's capacity for creating provocative, contemporary content. Despite critiques about AI's impact on journalism's core values like interpretation and creativity, the advancements in AI media are evident. This field explores the promises and challenges of AI-driven news, reshaping traditional journalism and prompting questions about accuracy, ethics, and the future role of human journalists. Noain-Sanchez (2022) sees AI integration in journalism as a double-edged sword: it enhances efficiency and journalistic capabilities. Still, it raises concerns over quality and media ethics, necessitating a balance between technological progress and preserving the fundamental values of journalism.

The evolving role of human journalists in the context of algorithm-driven journalism is becoming increasingly significant. As routine reporting tasks are automated (Winiarska-Brodowska & Feher 2025), a shift is seen in two directions. Firstly, combining AI with human journalism improves the creative writing process. Secondly, more analytical, investigative, and interpretive journalism emerges as routine tasks are automated. This evolution calls for rediscovering journalistic skills, the importance of AI and generative AI literacy—understanding AI's and GenAI's fundamentals, applications, and their societal impact (Noain-Sanchez 2022, Engelberg Center Live 2024). The synergy between AI and human journalistic skills transforms news production and the context of reporting.

The volume of related research is rapidly increasing, primarily aimed at understanding and predicting journalistic and entertaining practices, as emerging studies illustrate. The rise of AI in media was first extensively researched by the London School of Economic's POLIS think tank in 2019, finding that half of 71 organizations in 32 countries used AI in newsgathering, two-thirds in production, and half in distribution (Cooper 2020). Wang's research into Chinese news media revealed significant advancements in news gathering, writing, and distribution, particularly through robot writing, virtual news anchors, and *recommendation systems* where algorithms are designed to suggest items based on user preferences and behaviors (Wang 2021). Statista's analysis started to summarize how the media and entertainment sectors are pushing technological boundaries through AI integration (Guttman 2023).

The most pragmatic concepts, in this context, are process journalism and automated journalism. *Process journalism* emphasizes publishing news in its developing phases, allowing for real-time updates and corrections as new information emerges, ensuring accuracy throughout the reporting process. AI-driven process journalism involves AI in every production stage. Automated journalism also plays a crucial role in the news industry, transforming raw data into coherent news stories (Ford 2018) and showcasing AI's growing sophistication in emulating human journalistic skills. Despite cost-effectiveness, challenges like language complexities and mis/disinformation hinder automation in the news industry (Wang 2021, Broussard et al. 2019). Moreover, automated journalism is considered in relation to its potential to cause structural changes, alter power dynamics, and raise concerns about manipulation and censorship (Helberger et al. 2020). Accordingly, accuracy and transparency are repeatedly emphasized in academic and non-academic discussions in AI journalism, especially in the narratives about power dynamics. Its capability for evaluation is still under study (Jang et al. 2022), but as GenAI improves, the distinction between human and AI journalism may become less apparent.

Ananny and Finn (2020) propose a proactive approach to journalism through *anticipatory infrastructures*—systems designed to predict and respond to future needs or challenges using AI, data analytics, and sensors. These infrastructures shape social perceptions by anticipating events and modeling online attention.

This suggests that journalism's role extends beyond reporting current events to managing potential future scenarios, which is crucial for democratic engagement and participatory public discourse.

Synthetic-Hyper-Immersive Media Extensions

AI or generative mediatization and automation project future expansions that will develop and enhance the media experience. The concept of *Artificial, Synthetic, Hyper, and Immersive Media Extensions* (ASHIMEs) proposes a summarizing concept for the converging technologies with this, collectively employing computational content creation and organization methods. Synthetic media generates or alters audiovisual content, hypermedia structures dynamically link various content types, and immersive media integrates these techniques with AR/VR/XR and AI to create intuitive interfaces, interconnected contents, and interactive experiences (Lee et al. 2022, Milliere 2022, Signer et al. 2021).

In AI media, the convergence of these technologies results in highly personalized and adaptable content experiences, primarily via virtual influencers controlled by humans or synthetic influencers driven by AI. This extension is facilitated by emerging synthetic generators that scrape internet mosaics (texts, images, voices, and videos) and create algorithmically generated content. ASHIMEs is intensively learning from interface-based interactions and human behavior in non-real worlds. The critical question becomes the balance between real-world-based and ASHIMEs-based interactions and behaviors in different media services, with AI potentially playing a pivotal role in determining this balance.

The competition between hyperreality and the *Metaverse* concept envisions a synthetic future with several media technologies in convergence. Matsuda's hyperreality project (2017) represents a real-world moving space with always-on augmentation, whereas Meta aims to build a universal virtual environment for an omnipresent mediatized existence from business to education. The *hyperreality* concept, with a longer history and a more neutral or futuristic tone, contrasts with the intensively discussed metaverse with its controversial massive investments. As Pew Research Center and Elon University (2022) reveal split opinions among experts on the metaverse's future impact, the vast gap between the massive investments being funneled into it and the unforeseen consequences is concerning, primarily when these resources could instead be directed toward sustainability or promoting societal balance.

The more pragmatic media technology goals are also studied in related fields to understand upcoming trends. The first example of this is the *hologram-based service* option, such as the social robot "Hupo," which uses a three-dimensional image created with photographic projection to blend AI, AR, and anime, offering a unique form of anthropomorphism. This development holds significance for Human-Computer Interaction (HCI) and cultural studies, as it explores how tech-savvy users engage with AI technologies (Leo-Liu & Wu-Ouyang 2022). In journalism, AR's potential to enhance user engagement is being explored. The second example is focusing on AI. Aitamurto et al. (2022) studied AR's impact on users' sense of physical presence, knowledge gain, and perceptions of the authenticity of journalistic visuals. Their findings suggest AR induces a greater sense of physical presence but does not necessarily inform participants better or affect visual authenticity perceptions. The third example highlights gaming, where AI's application is exemplified in a game based on Raden Patah, using technologies like the Unity3D engine and Blender for 3D asset creation, and the Fuzzy Sugeno Algorithm for dynamic NPC behavior. This enhances the gaming experience and serves an educational purpose by incorporating local cultural content (Saputra 2017). The last example is *virtual and synthetic influencers*. Jin's study (2023) on AI-powered influencers highlights their impact on social media users, noting that lonely users prefer more human-like avatars, while those with high trait reactance respond negatively when their autonomy is threatened. In a compact summary of the generative AI-created influencers, brands are significantly provided with consistent, scandal-free content that can be automated and reliably managed (Ferraro et al. 2024). Even if they lack the attractive element of human vulnerability and mortality, GenAI makes them an ideal choice for AI-driven brand promotion in the long term.

These kinds of studies offer far more concrete insights into the development directions of AI media than the rampant debates about its endless future—though the latter's arguments may still be valid, as was mentioned above in the case of meta-investments. All demonstrate how ASHIMEs is being studied in various fields. However, the transformation to hyperrealistic and surrounding synthetic-immersive ecosystems is still in its early stages. The interconnection of ASHIMEs services could significantly change the media industry, but this requires intense and real-time interconnectivity of infrastructures, qualified data flow, human training, and legal-ethical adaptation.

Disinformation and Deepfakes

Disinformation and misinformation are widely studied across various disciplines and frequently examined within the context of political and marketing campaigns. *Disinformation* often involves deliberate fake information or falsehoods for manipulation or organized campaigns. In contrast, *misinformation* generally refers to unintentional errors in information. Regardless of intent, the outcome is the spread of inaccurate or misleading information, potentially causing crises in journalism (Snapp 2021), violent uprisings, and social instability (Opal 2021).

Online and *generative search* serves as a critical entry point for understanding the amplification of mis/disinformation. Aslett et al. (2024) found that online searching could increase belief in false information by around 20%, especially when low-quality information is prevalent. This challenges traditional media literacy approaches and underscores the need for empirically tested methods in media literacy programs, alongside improvements in search engine algorithms, in parallel with understanding the roles of fact-checking and skepticism (Hoes et al. 2024). Thus, AI services become critical hotspots for analyzing false information flows and identifying patterns for proactive intervention.

Deliberate mis/disinformation that is intended to mislead has been widely observed in organized campaigns, notably in countries like China (Jee 2019). This issue, although not new, has escalated with AI-driven media (Posetti & Matthews 2018). The psychological impact of disinformation is profound (Weikmann & Lecheler 2023), and the development of live deepfakes highlights the comprehensive and reality-altering potential of this influence (Mirksy et al. 2019). Addressing these challenges requires complex strategies, from the linguistic approach (Veszelszki 2017) to leveraging technologies like AI or blockchain for fake news detection and fact-checking (Lewis et al. 2019). It appears that AI technology is beginning to fight with itself, integrating further advanced technologies.

Deepfakes, a significant focus in media and communication studies, highlight credibility risks within synthetic media, necessitating cross-sector mitigation strategies (Guzman & Lewis 2024). While deepfakes use AI to manipulate human faces and voices, synthetic media covers a broader range of AI-generated content. This shapes our perception of digital and synthetic information, which is increasingly becoming our preferred source.

Deepfakes challenges core concepts of truth and authenticity. This technology creates highly realistic yet fabricated images, voices, and videos, posing challenges in a variety of contexts (Reuters 2021, de Seta 2021), including political disinformation, unethical manipulation (Hao 2021, Sanjeev 2023c), or idealistic hyperreality of branding. Their use in political propaganda and election interference threatens democratic processes (Diakopoulos & Johnson 2021), which necessitates robust measures and advancements in detection technologies while also recognizing the potential benefits of deepfakes in education, art, and therapy, provided ethical governance is in place (Danry et al. 2022).

Verification mechanisms or detection tools, like watermarks, filter tags, metadata labels, platform warnings, content flags, source badges, trust indicators, and algorithmic filters, have been suggested to address AI-generated manipulation, which involves using AI tools to shape or distort information for targeted outcomes (Shane et al. 2021). This approach is vital for responsible media technology and requires interdisciplinary efforts (Trattner et al. 2022), implying a comprehensive, multi-stakeholder collaboration. Deepfake technology and AI-driven disinformation represent a new era of digital deception, with profound implications for politics, media, and public perception. As these fabrications become more sophisticated, discerning reality from manipulation becomes increasingly challenging, necessitating a multi-faceted approach that includes advanced detection methods, legal frameworks, public education, and global cooperation. The broader implications of these challenges are further explored in the chapters on "Ethics" and "Policy."

In the wake of the post-truth era, where facts hold less weight in public discourse, Asta Zelenkauskaite's *post-public concept* in *Creating Chaos Online* shows how human and nonhuman actors, like bots, disrupt traditional discussions by mixing real and manipulated content. Tactics such as *whataboutism*, a rhetorical deflection technique using unrelated issues to avoid criticism, further fragment public debate, complicating democratic clarity (Zelenkauskaite 2022). This disruption also impacts areas like scientific disinformation (Falyuna 2022), which is crucial to study as it erodes public trust in evidence-based decisions and spreads through media, as seen during the recent pandemic.

Considering the potential impacts of deepfakes and disinformation in the context of post-public, detection tools, legal-ethical frameworks, and AI literacy alone are insufficient to maintain information integrity. All of these must be scaled up to support AI media and generative AI, creating responsible AI-driven media services by design. Organizations are increasingly adopting proactive education and preventive measures over punitive responses to ensure ethical AI usage (Dabis & Csaki 2024).

As our recent research findings emphasize, comprehensive generative AI and AI media literacy is required, integrating technology education from early childhood on through higher education, and involving stakeholders such as businesses, educators, and parents to foster critical, ethical technology practices across all generations (Feher 2024c).

Recommendation Media and Mind Control

Social media platforms have evolved and transformed into recommendation-driven media hubs. Initially, interconnectedness and interactivity were the focal points of social media (Li et al. 2021). However, as Mignano (2022) highlights, there has been a shift from traditional social media to recommendation media. Platforms such as TikTok and YouTube now prioritize algorithmic feeds over social connections, reflecting a transition aimed at optimizing content distribution for engagement. This shift both presents challenges to traditional social media models, and potential solutions to counter low-quality content proliferation. Rourke (2021) notes that business models of these platforms increasingly rely on AI-automated decision-making and personalized customer experiences, ranging from ad-dependent publishers to e-commerce and the media industry. These business models showcase diverse approaches to media monetization and user engagement, where audience content recommendations become not just a technological feature but a fundamental business competency (Chan-Olmsted 2019). Lewis (2023) metaphorically compares these platform-driven business models to coffee houses, where users are not just customers but also products in a surveillance system, highlighting the invasive nature of data collection and personalized advertising in AI-supported platforms.

AI-driven recommendation systems use hyperpersonalization, predictive analytics, and chatbots to enhance user experience by tailoring content and interactions based on individual behavior. In this process, *hyperpersonalization* adapts content in real time, predictive analytics anticipate user preferences for future interactions, and chatbots offer seamless communication, making the system more engaging and responsive. As these technologies grow, concerns about their direct impact on human behavior increase, with many powerful actors still skeptical about major investments in AI media. For instance, Sanjeev discusses the alignment problem in recommendation algorithms, citing YouTube's role in radicalizing the Christchurch shooter (2023b). While AI-driven social platforms are profitable, this focus on maximizing views or viewing time can disturb users, revealing the ethical challenges in these business models. Similar considerations are critical in AI-powered advertising, especially given public concerns about manipulation and security vulnerabilities (Wong et al. 2023). In an ideal world, hyperpersonalization strategies are aligned with business policies, focusing on customer engagement while considering societal impacts. However, in the real world, vulnerable groups can be adversely affected by targeting and recommendation systems (Bol et al. 2020), or self-endorsements by intelligent personal assistants can have unintended psychological impacts on brand attitudes and purchase intentions (Hamilton et al. 2021). Business-driven services and content providers must work with deeper insights to mitigate these impacts.

The potential for *mind control* through AI media and Generative AI is significant, as these technologies can manipulate thoughts and behaviors by integrating AI, neural interfaces, and real-time synthetic content. By combining Neuralink and AI-driven content, these systems could directly influence neural activity, shaping societal perceptions or promoting cultural homogenization. As such transformative technologies converge, there is a growing demand to protect users from manipulation and disinformation (Rosenberg 2022). Based on their study of TikTok users, Siles et al. (2022) suggest introducing algorithmic awareness for users, considering how changing recommendations affect user experiences, perceptions, and interactions with the app. Their findings emphasize the importance of investigating frustrations with algorithmic training and content over-saturation. While AI-driven content moderation can reduce unexpected effects (Ali Abbasi et al. 2022), it cannot fully replace human judgment due to limitations in contextual and cultural nuances.

Dangerous Optimism

The human factor plays a crucial role in shaping and envisioning the future of AI, as emphasized in the "Introduction" chapter. Our recent study reveals that AI media and info-communication experts from all world regions primarily focus on AI trust and strategy, oscillating between their beliefs and uncertainties. In their "AI trust model," interestingly, the balance often tips toward hopes and beliefs, leading to *dangerous optimism* or overoptimism. Therefore, they tend to pass responsibility onto future generations, technology, and nature-inspired solutions. This over-optimist and responsibility-deflecting attitude can be hazardous if these experts are shaping the future of AI media and information communication (Feher et al. 2024). This trend can align with the overestimation discussed in the chapter on generative AI (Galloway 2023, Brynjolfsson 2022).

Optimist narratives are influenced by two main factors: time and context. Regarding time, our backcasting research involved AI experts envisioning a distant, optimal future and then working backward to plot a realistic path to reach this future. This method challenged them to balance pragmatic and skeptical views (Koves et al. 2024). In terms of context, our other research into investors and developers of AI revealed a realistic and positive vision when discussing decision-making and calculation-based achievements. However, when the conversation shifted to media and behavior manipulation, fears about, and risks for, the future surfaced, leading to extremely negative visions (Feher & Veres 2023). Despite their exaggerated fears, continued investment in AI technology indicates that the concerns are more critical than emotional, considering the context of profit maximization.

However, through our studies, we have also observed that the selection of research methods and concepts, such as horizon scanning (Feher et al. 2025), backcasting (Koves et al. 2024), and the sociology of expectations (Borup et al. 2006), significantly influences the ratio of optimism versus skepticism in research outcomes, particularly in critical areas like public trust and societal integrity. Therefore, by carefully selecting methods, time frames, and contexts, researchers can better anticipate challenges and guide responsible innovation, ensuring that the rapid advancement of AI aligns with societal values and ethical principles

It is also essential to understand how visions shape both the time frame and context of the future of AI in media and communication. Ashworth (2023) advocates a human-centered approach, noting that communications leaders know AI adoption challenges like security, privacy, and disinformation. Accordingly, they need to develop new mindsets and skills for strategic thinking. The psychological impact of deepfakes, along with the challenges faced by various social, age, or gender groups (Yadlin-Segal & Oppenheim 2021, Jabłońska & Zajdel 2020), calls for pragmatism in the AI media.

The concept of *infinite media* indicates a paradigm shift in the media industry toward limitless content production and consumption, particularly in Hollywood, and on AI-driven platforms like social media and streaming services (Echauri 2023). Generative AI could further accelerate these trends, pushing us toward an era of limitless AI media. However, excessive optimism in this direction could lead to chaos or a new balance in the media and news industry. Thus, understanding the double-edged sword of the related technologies is vital to addressing ethical, societal, and psychological challenges (CPA Canada 2022), especially in long-term visions focusing on human-centered values, critical thinking, and preventive actions.

Impact Project

Bloomberg's Cyborg represents a landmark in applying Generative AI in financial journalism. Specifically designed to process vast quantities of complex financial data, Cyborg uses advanced Natural Language Processing (NLP) to extract key insights from quarterly earnings reports and other intricate financial documents. The system can rapidly transform this data into concise, real-time news summaries and headlines, a precious asset during critical periods such as earnings season, where the speed of information delivery can influence market movements. By automating much of the data extraction process, Cyborg enables Bloomberg to break financial stories faster than competitors, reinforcing its status as a leading source of reliable financial news and analysis.

This initiative is not merely about efficiency but about transforming how financial journalism operates. Cyborg's ability to process large volumes of data faster than any human team could highlight a shift toward automated reporting, where speed is paramount. Cyborg's rapid data processing offers a significant competitive advantage in an industry driven by immediacy, setting new standards for gathering and disseminating news. Yet, the system still relies on human journalists for tasks that require contextual understanding, deeper analysis, and critical interpretation, underscoring a collaborative dynamic between AI and human expertise.

Bloomberg's Cyborg initiative is particularly proactive in aligning itself with the algorithm-driven financial markets, which depend heavily on speed and precision. In an environment where millisecond-level financial transactions can influence market movements, Cyborg's ability to deliver real-time financial news creates a powerful feedback loop, directly impacting stock exchanges and financial decision-making. This symbiotic relationship between AI-driven news and AI-driven trading accelerates the dynamics of the financial and media industries and reverberates through the global-regional economies. Decisions influenced by these rapid news cycles can cascade into business strategy and even shape geopolitical outcomes. Bloomberg further leverages this advantage through its cross-platform integration. Pivotal GenAI financial news—generated by AI but contextualized by human journalists—is broadcast on their key platforms and TV channels and is also delivered to screens of prestigious business hotel rooms. This ensures critical decision-makers receive up-to-the-minute insights, reinforcing Bloomberg's role in shaping the business and political ecosystem.

However, such a system's broader societal and economic impacts are far-reaching in different industries. In the media industry, tools like Cyborg highlight the growing shift toward AI-driven newsrooms, which demand new skills from human journalists, such as proficiency in AI oversight and data interpretation. As AI systems take over routine reporting tasks, human journalists can focus on more analytical and investigative work, potentially elevating the overall quality of journalism—provided they receive adequate support and opportunities for upskilling in GenAI. Yet, there are concerns about job displacement and the reduced role of entry-level journalists traditionally tasked with routine reporting.

The integration of GenAI is already sparking tension across industries. While AI tools can churn out content, streamline production, and automate processes, the real question is how much human ingenuity and creativity we are willing to sacrifice—and at what cost. The question arises whether AI should be permitted to replace critical aspects of human contribution or if doing so risks undermining these industries' intrinsic value. The potential for collaboration between AI and humans presents opportunities for increased efficiency and innovation, yet it also raises concerns about preserving human involvement's ethical and creative dimensions. Finding a balance between leveraging technological advancements and safeguarding human insight may be crucial in managing these evolving dynamics, though the path forward holds several uncertainties.

Reflective Art

Integrating human and machine-generated content in art unlocks creative possibilities and prompts critical questions about authenticity and AI's future impact on socio-cultural perceptions. AI's capacity to augment art forms into new options is evident. However, AI algorithms can generate unconventional scales that challenge traditional compositional norms in artwork. In the visual arts, AI creates surreal compositions, while in storytelling, it constructs narratives that defy conventional structures, pushing the boundaries of art and offering insights into non-human interpretations of the world (Vincent 2021).

Experimental art increasingly utilizes AI media in various ways. One general approach explores the differences between traditional and AI-driven art methods. Rourke (2021) highlights hyperpersonalized content creation as a prime example of AI's transformative impact, where algorithms tailor media experiences to individual tastes far beyond the capabilities of traditional methods. This hyperpersonalization leads to the creation of extremely niche content. Rosenberg (2022) discusses the transgressive use of AI in art, where artists employ AI to challenge societal norms and artistic conventions, producing provocative artworks that stimulate discussions about technology's role in art and society. Another area is the exploration of unconventional artistic collaborations enabled by AI across various mediums, encouraging unique and experimental art forms (Jin 2023).

Reflective art of AI media, accordingly, focuses more on AI-driven art's societal impact and also provides noteworthy examples. Generated non-linear and fragmented narratives in writing, journalism, or filmmaking challenge storytelling norms (Kruger 2023), resulting in extreme presentations. Sharp and controversial messages, often more radical than those typically produced by human artists, are created to encompass social and political commentaries (Hao 2021). Diakopoulos and Johnson (2021) discuss ethical extremes in AI-generated content, noting that deepfakes and synthetic media blur the lines between reality and fiction, departing from traditional reality checks.

The Dead End Gallery (2024) in Amsterdam is a significant milestone in accepting AI-generated art. As the first gallery to exhibit works from the fictional AI artist Irisa Nova, it challenges traditional artistry and sparks debate on AI-generated art's value and authenticity, resonating with themes of generative mediatization and synthetic extension.

The AI Surrealism Exhibition (Exquisite Workers 2023) presented a symbolic step forward on this track, showcasing the world's largest AI art exhibition in 2023. This event emphasized AI's dynamic and provocative role in art, aligning with themes of optimism. It opened discussions about art's future direction and its industry norms.

The Charismatic Project, backed by Innovate UK, is a groundbreaking AI initiative aimed at transforming media and storytelling in film and television (UAL 2024). Led by Charismatic.ai, Channel 4, and UAL, it leverages generative AI to support underrepresented creators and established producers, helping them use AI-driven tools for content creation, including storyboards and scenes. By focusing on AI-enhanced media innovation, the project seeks to democratize access to technology, encouraging diversity in media storytelling while exploring the ethical applications of generative AI across creative industries.

These contemporary AI-generated artworks offer a glimpse into AI media art's evolving field, showcasing its promising potential and complex challenges. They capture how technology and creativity work together, allowing AI's ability to hyperpersonalize art and transform how audiences experience it, creating unique and individualized experiences. They also highlight *algorithmic awareness*, understanding how these generative and mediatized technologies function and shape adaptivity with broader societal consequences.

Conclusion

AI media involves integrating and adapting AI into media systems, leading to AI mediatization in an interplay of generative mediatization. The applications of AI in media, ranging from news to entertainment, are defined by content creation, automation in media production, hyperpersonalization, and distribution. The exploration of generative AI, in particular, highlights its vital role in content transformation, robot journalism, deepfakes and mis/disinformation detection, synthetic-hyper-immersive media extension enhancement, and the responsibility of recommendation systems. Accordingly, it is essential to recognize that a human-centered approach remains vital in shaping the future of AI in the media sector. We, however, must carefully select research methods involving human participants to avoid over-optimism and bias, ensuring that skepticism and critical thinking are always addressed. Likewise, dystopian scenarios can be addressed through well-designed research designs for pragmatic solutions. The ultimate goal is to shape media content and its dissemination to influence public perception or agenda.

Future research could explore how AI media and metamedia challenge user trust and interaction by transforming engagement and content creation. The evolving AI-to-AI loop offers new insights into how user experiences are extended and manipulated, especially in the spread of mis/disinformation and deepfakes. Ethical concerns around mind control and infinite media, as well as the role of algorithmic awareness and neuralink technologies, demand closer scrutiny. The ultimate question, therefore, is not whether AI will shape the future of media but whether society is ready to (partly) delegate control of its own narratives.

Scholarly Debates

For Cultural and Social Sciences

- How do deepfakes challenge traditional concepts of truth, authenticity, and credibility?
- What are the socio-cultural impacts of using AI for content moderation and mis/disinformation detection?
- How does automated journalism transform power and control over societal narratives?
- How is AI-driven media expected to evolve in the future while balancing optimism and skepticism about its broader impact?

For Science, Technology, and Engineering

- What technical risks does hyperpersonalization pose, and how can they be managed?
- How will AI's content generation transform media systems and workflows?
- How do AI-driven recommendation systems impact user engagement and optimize media consumption through data-driven approaches?
- How can AI systems be designed to balance automation with human oversight and creativity?

For Business, Policy, and Innovation

- What regulations are necessary to prevent AI-driven mis/disinformation and deepfakes?
- How can policies ensure transparency and explainability in AI media to protect public trust and societal integrity?
- How can businesses use AI in media ethically while complying with laws on hyperpersonalization and recommendation systems?
- What strategies can businesses implement to mitigate the risks of over-optimism in AI media?

Strategic Takeaways

For Academia

- AI Detection Research: Explore advanced AI detection tools to identify deepfakes and verify media authenticity.

- Cross-Disciplinary Collaboration: Establish interdisciplinary research that integrates AI with humanities and social sciences to gain a comprehensive understanding of AI's broader societal impact.
- Future of AI Journalism: Investigate the evolving role of AI in journalism, focusing on news accuracy, integrity, and reliability.
- Future of AI Media: Utilize strategic foresight through a robust research design to balance optimism and skepticism in shaping AI media's future.

For Industry

- AI-Driven Media Creation: Explore process journalism and new storytelling formats to generate innovative content.
- Engage and Moderate: Improve user engagement by implementing content moderation and fake media detection.
- ASHIMEs Readiness: Explore synthetic, hyperrealistic, and immersive media environments to advance business strategies.
- Explainable AI in Media: Focus on ethics and transparency in media production to ensure AI models are understandable and align with responsible practices.

For Policymaking

- AI Literacy and Public Awareness: Promote policies to improve AI literacy, and raise public awareness of AI's role in media, including its potential for mind control.
- Intellectual Property Reform: Update intellectual property laws to address the challenges AI-driven content creation poses.
- Global AI Governance: Encourage international collaboration to develop unified standards and policies for the era of AI-driven media.
- Disinformation and Recommendation Systems: Create policies to combat disinformation and manage AI-driven recommendation systems influencing human behavior.

4

UNCERTAINTIES

Defining Uncertainties

In generative AI and synthetic media, uncertainties refer to the unpredictable outcomes of integrating these emerging technologies into society. These uncertainties primarily stem from the complexity of AI innovation, ethical challenges, mis/disinformation issues, and the development of emotional technologies. While uncertainties can foster creativity and innovation, they disrupt existing business models and raise concerns about sustainability and societal impact. Effectively managing these risks requires carefully balancing potential rewards and unforeseen consequences. Uncertainties blur the lines between utopian and dystopian futures, influencing societal trust, business practices, and AI's impact on environmental sustainability. Artistic reflections on these uncertainties provide a creative lens to understand AI's influence on societal trust and the nature of creativity

Introduction

The rise of generative AI and AI-driven media forces us to understand and reevaluate how we handle uncertainties in a world shaped by rapid innovation. Unlike traditional risks, these technologies' uncertainties are more challenging to predict and control. Particularly challenging to anticipate is their impact on societal perception, social-environmental sustainability, and problem-solving. Generative AI's ability to blur the boundaries between truth and fabrication or originality and authenticity poses immediate challenges, where public trust and accurate information are already under strain. Addressing these uncertainties requires innovative solutions which balance technological advancement with societal responsibility and ensure that innovation does not come at the cost of deepening inequalities or environmental harm.

DOI: 10.4324/9781003591023-5

The nature of uncertainty in generative technologies is complex, intertwining risks that can be quantified with uncertainties that remain elusive and non-visible. This dichotomy challenges our conventional approaches to problem-solving and innovation, urging a shift toward more sustainable and emotionally intelligent strategies. In AI, particularly with technologies like deepfakes and algorithm-driven content creation, the ability to predict future outcomes and understand societal perceptions becomes increasingly critical. These technologies hold the power to reshape the geopolitics of information and the global economy of data, highlighting the urgent need for a nuanced understanding of AI's potential and pitfalls.

This chapter explores the root causes of uncertainty in generative AI and AI media, examining how these uncertainties intersect with societal, business, and environmental challenges. As AI transforms industries, its reliance on content mining and neuromorphic computing introduces instability in risk assessment while raising concerns over data ethics and emotional dependency. Thus, a comprehensive uncertainty approach is proposed in deploying generative technologies. Through this lens, the chapter underscores how multifactorial uncertainty considerations can balance the essential aspects of a society or business.

The Nature of Uncertainty

The "nature of *uncertainty*" refers to the unpredictable and often elusive qualities defining our interaction with human experience and generative technologies. Uncertainty embodies complex narratives, filled with intricate plots and layered themes, where outcomes cannot be easily predicted. These narratives challenge our understanding of reality, demanding a deeper engagement with how humans and machines confront the challenging uncertainties.

The discourse on uncertainty management, as articulated by Kramer (2018), serves as a cornerstone for understanding how individuals and organizations construct meaning amidst ambiguity. *Uncertainty management* is defined as a strategy to handle and reduce the unpredictability and indeterminacy in AI systems and their outputs. This intellectual exploration, stretching from personal introspection to organizational strategy or beyond to geopolitics, underscores the quintessential human quest for clarity in the dynamics of the unforeseen.

The juxtaposition of human and artificial responses to uncertainty unveils a profound dichotomy. Humans, trapped in the web of their consciousness, often confront uncertainty with a palpable sense of hesitation, their assertions laced with the *vulnerability* of doubt—a flaw or weakness in a system that can be exploited to compromise its security, functionality, and reliability. In stark contrast, artificial intelligence, such as ChatGPT, DeepSeek, or other generative AI services, traverses uncertainties with an unsettling equanimity, delivering assertions devoid of the hesitance that typifies human judgment Kruger (2023), reinforcing AI fatalism (AI Action Summit 2025) over human agency.

This divergence in confronting uncertainty beckons a deeper inquiry into the essence of intelligence and the pivotal role uncertainty plays within it. Artificial intelligence, guided by the principles of utility maximization and reward systems as delineated by Russell and Norvig (2020), adopts a markedly different stance toward uncertainty, predicated on probabilistic outcomes and a relentless pursuit of optimal rewards. While formidable in its computational prowess, this mechanical approach occasionally falters, betraying the limits of encoding the depth of human uncertainty management into algorithms.

In the case of foundational models, the large pre-trained systems can handle multiple tasks via *neuromorphic computing*, brain-inspired hardware designed for efficient processing—without consciousness of uncertainties. As Nguyen et al. (2023) highlighted, these technologies herald a new frontier, potentially bridging the gap between AI's structured logic and the fluid ambiguity of human cognition. With a promise of enhanced processing capabilities and nuanced data interpretation, these pioneering technologies can trigger a future where human intuition or sense of uncertainties can also become synthetic.

In contrast, the challenges of uncertainty encourage a rethinking of uncertainty not as an adversary to overcome but as a fountain of creativity, innovation, and profound existential insight (Feher 2022a, 2022b). This perspective challenges the negative uncertainty narratives, proposing a paradigm where the unknown becomes an opportunity for intellectual and artistic exploration. Giving machines an understanding of uncertainty mirrors humanity's ongoing journey of self-awareness. Uncertainty constantly reshapes how we search for meaning, face the mirror of synthetic content, and deal with the unknown.

Uncertainty vs. Risk

Exploring the labyrinth of transformative technologies, we confront the intricate concepts of uncertainty and risk. As articulated by Sibert (2020), *uncertainty* embodies the myriad unknowns that pervade autonomous systems, ranging from normative ambiguity to the limitations imposed by bounded rationality. This enigmatic aspect of AI challenges us to devise robust frameworks for meaningful human control, acknowledging that the essence of uncertainty lies in its inherent unpredictability and the indeterminate nature of outcomes. The evolving performance of generative models, observed in Stanford's study, exemplifies AI capabilities' dynamic and unpredictable trajectory, underscoring the fluidity and elusiveness of certainty within artificial intelligence (Keary 2023).

Risk, by contrast, is the quantifiable and *measurable probability* with consequences of adverse outcomes that can be systematically assessed and managed. This domain invites a more structured approach to anticipating and mitigating the potential pitfalls associated with AI deployment, from ethical dilemmas to operational vulnerabilities. As leaders like Sam Altman highlight, the discourse on the possible dangers of AI technologies underscores the urgent need for vigilant oversight. Responsible stewardship is essential to harness the transformative potential of AI while mitigating risks. These risks include the possibility of AI enabling disinformation or cyber threats (Helmore 2023).

Orf's (2023) metaphor of the *ouroboros*—the serpent eating its own tail—illustrates the risks inherent in generative AI systems. As these systems increasingly process and feed on their synthetic outputs, the risk of degradation becomes more pronounced. This self-referential loop threatens to reduce the diversity and richness of the content AI produces, leading to a narrowing of perspectives and potential biases in the information it generates. The ouroboros metaphor underscores how generative AI could trap itself in a cycle of diminished quality without critical oversight, thereby exacerbating the uncertainties that already permeate AI-driven technologies. This phenomenon reflects a broader philosophical concern about the integrity of knowledge production and may distort rather than clarify reality.

Mitigation via data science and error management emerges as a critical domain in uncertainty, where the collaborative interplay between domain experts and data scientists illuminates the path to reducing risks. In line with this, Lin and Jackson (2023) highlight the transformative potential of viewing errors not as failures but as catalysts for innovation and learning, promoting a culture of resilience and continuous improvement that is paramount in the ever-evolving ecosystem of AI.

The intricate relationship between uncertainty and risk underscores a dynamic tension between foreseeable outcomes and unpredictable events. Two strategic foresight models illustrate this dichotomy. *Wild card* approaches anticipate rare, high-impact disruptions that require swift adaptation, such as an AI breakthrough in deepfake generation impacting media credibility (Feher et al. 2025). *Black swan* scenarios address unpredictable, extreme events with lasting consequences, such as a global AI-driven data breach altering privacy standards (Taleb 2005). These two frameworks reveal the unique nature of uncertainties, which resist precise quantification, prompting a deeper inquiry into generative AI's role in society. Unlike risks, which can be assessed and mitigated, these uncertainties call for flexible strategies to address an evolving technological landscape.

Innovation and Metareasoning

AI-driven innovation is filled with hype cycles and uncertainties. The concept of metareasoning helps us manage this ambiguity by reflecting on cognitive processes in problem-solving activities. As Sibert (2020) suggests, metareasoning equips AI systems with the capacity for nuanced decision-making, allowing them to achieve greater control over complex tasks. By reasoning to solve problems, AI can adopt more rational and adaptive approaches to optimal computations, even when outcomes are unclear. This process enhances AI's ability to manage uncertainty while supporting principled, normative approaches to decision-making. However, the path to innovation is fraught with unexpected costs, as noted by IBM (2023), emphasizing the financial implications accompanying the adoption of generative AI technologies. These costs represent a tangible barrier to innovation, particularly for entities unable to bear the financial burden, potentially stifling diversity and inclusivity in AI. Therefore, the degree of uncertainty is not only an issue of innovation but also of fragile competitiveness.

Thus, beyond the substantial benefits and advancements (Czerkawski et al. 2023), transformative technologies also challenge the job market, authenticity, and AI security. The hyperpersonalization capabilities of LLMs, as highlighted by Arguedas and Simon (2023), exemplify this duality of AI innovation well. Although generative AI startups can improve accessibility, user experience, or effectiveness by 30–60% (Springs 2024), they can also transform knowledge production or echo the mis/disinformation noise. This dilemma underscores the ethical and societal implications that accompany technological advancements, challenging developers to balance innovation with the broader impacts on society.

Araujo et al. (2020) shed light on the comparative effectiveness of AI in decision-making, where AI's judgments are often seen as on par with or superior to those of human experts in specific contexts. The rapid expansion of GenAI in decision-making and content creation has significantly intensified this sensation. This perception, driven by the machine heuristic and algorithmic appreciation, suggests a growing trust in AI's problem-solving capabilities. However, this trust must be tempered with caution, as the potential for AI to spread mis/disinformation due to a false sense of confidence, as Kruger (2023) points out, represents a significant risk. This false confidence can undermine the fact-checking attitude and deep critical thinking.

Accordingly, uncertainty management plays a crucial role in driving innovation. Occurring mainly through metareasoning, uncertainty management equips AI systems with the capacity for nuanced decision-making. This approach enables generative AI and AI-driven media technologies to manage complex tasks more effectively, allowing for greater control in unpredictable environments. Also, with metareasoning, businesses can better manage market concentration risks and technological monopolization trends. This strategic foresight ensures that advancements in generative AI contribute to innovation and equitable and sustainable progress rather than exacerbating global disparities or unforeseen societal-economic consequences.

Challenged Business Models

Generative AI and AI-driven media are fundamentally transforming business models, merging long-standing frameworks with rapidly evolving ones. This convergence highlights the tension between established digital strategies and the fast-paced innovation AI brings. Permanent models like *subscription-based services* remain crucial, yet new alternatives, such as content mining and hyperpersonalized AI media, are rapidly emerging.

The most permanent and highly applied business models in the digital or AI ecosystem are subscription and platform licensing. AI tools and platforms offer services, content, and information through recurring fees, allowing the integration of AI operations. This model provides stable revenue streams but demands continuous innovation to retain subscribers. The pressure to constantly evolve can strain resources, leading to potential sustainability issues (Riedl 2019). Despite its stability, the paradox of needing perpetual innovation to remain competitive can undermine long-term goals and hinder the overall sustainability of such business models.

Another prominent model is content creation and *customization*. AI tools allow for the automated production of various media formats, such as articles, videos, music, and design, allowing businesses to personalize user content. AI-driven content creation and customization offer a significant advantage in scalability, allowing for rapid media production at speeds far beyond human capabilities. However, McStay (2018) argues that this efficiency often comes at the cost of emotional depth and nuance, which are critical for human interpretation. Moreover, the model faces ethical challenges, as AI may prioritize efficiency over creativity and diversity. This raises concerns about homogenization and the potential for AI-generated content to contribute to disinformation, complicating the moral responsibilities of businesses. Accordingly, these challenges increase the costs of developing business models when societal and human principles are embedded into AI design. Despite this, that design appears on different levels as social fine-tuning improves customer engagement and societal ties.

Generative AI's role in advertising is another area of business transformation. Hyperpersonalized advertising has become a dominant model, with AI tools analyzing vast datasets to deliver highly targeted advertisements. While this allows companies to optimize their marketing strategies and enhance precision, it raises privacy and data manipulation concerns. Zuboff (2019) emphasizes that AI-driven marketing prioritizes profit over consumer trust, potentially eroding brand integrity in the long run. As personalization becomes more precise, the ethical risks of data exploitation become more significant, forcing companies to confront the broader implications of AI's role in consumer manipulation.

A more novel business model is the rise of synthetic influencers and AI-driven entertainment (see more details about synthetic influencers in the "AI media" chapter). AI-generated characters, increasingly prevalent in synthetic or virtual spaces, engage audiences through social media and gaming platforms. These synthetic entities generate revenue through sponsorships, brand partnerships, and in-game purchases, even if several individual and societal challenges surround them (Gunkel 2018). These characters often perpetuate unrealistic standards, distorting user expectations and contributing to problematic societal norms or uncontrolled subcultures. Moreover, issues of accountability arise when interactions with AI-driven characters blur the lines of responsibility, making it challenging to address the consequences of AI's actions.

In addition to these models, content mining has emerged as a critical practice, particularly in training AI models, regenerating tons of previous content. *Content mining* is extracting and processing vast amounts of digital data, often without proper consent or compensation to content creators, to train generative AI systems. The Informa-Microsoft deal, for instance, allowed Microsoft to access academic content for AI training without informing or compensating authors (Potter 2024). Similarly, the agreement between News Corp and OpenAI, valued at over $250 million, underscores the growing commercial value of content mining. These all raise serious questions about intellectual property rights and the ethics of knowledge production or authenticity (Bruell et al. 2024). These cases highlight the tension between innovation and fairness, as content creators increasingly find themselves left out of profit-sharing agreements despite their work being integral to AI's development.

As AI business models continue to evolve, alternative approaches are gaining traction. Accordingly, Floridi (2016) advocates for integrated ethical frameworks in the case of AI and human collaborative creativity. This reflects a broader shift toward business models prioritizing human-AI collaboration rather than replacing human labor. In this AI-driven creator economy, individuals and institutes use AI-driven content creation to blend human ingenuity with technology's capabilities. While still emerging, this model offers a more sustainable approach to balancing innovation with ethical considerations.

Societal-Environmental Sustainability

Human beings exist within a complex web of technology and the natural environment, with shared ethical, ecological, and societal consequences. Transformative generative AI and AI-driven media result in massive energy consumption, inherent data biases, and labor market disruptions. Paradoxically, this process intensifies technological use even further. Those who have mastered the technology seek to expand their competitive edge, primarily with an optimist approach. Others grow increasingly concerned about potential resource constraints caused by its broader, widespread adoption. Both camps tend to assess the sustainability of infrastructure and data traffic in a narrower scope, often overlooking broader social impacts. Consequently, discussions surrounding these issues frequently downplay the societal consequences, focusing instead on technical and environmental aspects.

Socio-environmental sustainability, instead, is proposed as a balanced approach to force future-oriented and trusted generative AI and AI media. In this case, environmental and societal sustainability are in the two scale spans. Environmental sustainability emphasizes minimizing the carbon footprint, energy consumption, and resource demands required by AI models. Social sustainability focuses on fairness, inclusivity, and reducing societal harms like job displacement or mis/disinformation. If these two levels of sustainability are balanced, they can effectively contribute to a resilient ecosystem that integrates societal well-being, environmental health, and technological progress. This balance fosters a responsible approach to development, ensuring that human, ecological, and technological systems can adapt to each other and thrive together over time. The societal-environmental balance is a strategic imperative to ensure AI's benefits do not come at the cost of societal well-being and ecological balance. The responsible development of AI, including generative models and synthetic media impact, must center on these dual imperatives of societal well-being and environmental sustainability.

Regarding environmental sustainability, one of the most urgent challenges posed by AI is its immense energy consumption. The rapid expansion of AI-driven media and generative models has caused a sharp rise in demand for computational resources, significantly impacting global energy use. Unlike smaller-scale applications such as smartphones, AI models used in high-demand industries—such as natural language processing, image generation, and other resource-heavy operations—mirror the energy consumption levels of the top CO2-emitting industries. Optimizing energy storage and consumption in AI infrastructure is a significant hurdle that must be addressed to prevent worsening environmental impacts. Nishant et al. (2020) argue that AI's heavy reliance on vast datasets and unpredictable outcomes—especially in ecological contexts—raises serious concerns about its ability to mitigate or exacerbate climate change. The energy-intensive nature of generative AI highlights the need for critical interventions. AI systems consume vast amounts of energy, comparable to industrial levels, which demands the urgent adoption of renewable energy, more efficient algorithms, or further optimization methods (Hodgkinson & Jackson 2023, Heaven 2023b, Feher 2023a, 2023b). Industry-wide efforts are needed to curb AI's environmental footprint, mirroring strategies already in top CO2-emitting sectors or finding solutions through AI innovations. These actions are crucial to reducing the ecological costs associated with the rapid growth of AI technologies.

The societal implications of generative and AI media are also significant. Generative AI's rapid automation of labor-intensive tasks, particularly in content creation, threatens job stability, either by displacing human workers or by necessitating reskilling to adapt to new roles, thus transforming multiple industries intensively. The acceleration of these changes introduces several uncertainties, requiring careful management to balance innovation with the societal need for sustainable employment and workforce adaptation. The potential for AI-driven disinformation or misuse of deepfakes also raises significant concerns about AI's impact on social cohesion and public trust. A symbolic case involved Indian journalist Rana Ayyub, who was targeted with AI-generated deepfake pornography in an online smear campaign. Her face was superimposed onto explicit content as part of an attempt to discredit her and intimidate her professionally. This form of AI misuse highlights grave concerns about cyberbullying and the erosion of personal privacy and safety. The societal consequence is uncertainty in trustworthiness, where public figures and individuals are increasingly vulnerable to falsified content that can undermine their credibility and disrupt social stability. Moreover, as Goode (2023) discusses, AI introduces new ethical complexities through phenomena like the *memory effect*. This is AI's ability to subtly alter personal memories and perceptions via deepfakes, blurring the distinction between authentic experiences and AI-generated content and threatening societal trust in shared histories. These disruptions and AI's growing role in shaping digital environments underscore the critical need for responsibilities in robust societal sustainability frameworks. Without addressing these uncertainties, generative AI risks exacerbating inequality, eroding trust, and reshaping societal norms unpredictably. Balancing AI-driven societal cohesion should be reconsidered accordingly for societal sustainability.

Addressing these challenges requires innovative approaches like *climate literacy* and responsible data circularity to minimize harm while enhancing long-term ecological and social resilience. Luccioni et al. (2024) underscore the sustainability challenges associated with multipurpose generative AI systems, noting that these models are significantly more expensive and resource-intensive than task-specific counterparts, necessitating careful management for responsible growth.

Emotional Intelligence

The concept of emotion AI, or affective computing, is a technology that intersects with the most intimate aspects of human experience: our emotions. This emerging field promises to significantly transform how we interact with machines through generative AI and AI-driven media. Yet, it also raises serious concerns about the nature of these interactions and their implications for societal integrity, emotional intelligence, and individual well-being.

Emotion AI lies at the heart of understanding and interpreting human emotions through advanced algorithms and machine learning techniques. As Purdy (2019) notes, businesses and organizations increasingly adopt emotion AI technologies to decode facial expressions, analyze voice patterns, and monitor eye movements. These systems aim to capture the emotional reactions of employees and consumers in real time. In line with this, their services offer opportunities to improve work efficiency and customer satisfaction while also developing products that resonate deeply with users' habits, even reaching levels of personal connection and intimacy. However, the subjective nature of emotions introduces challenges, especially the risk of bias, error, lack of empathy, or misinterpretation of the context in emotional assessment. Emotional analysis technologies, if not carefully designed, could inadvertently perpetuate stereotypes or reinforce existing prejudices, amplifying societal inequities and undermining the ethical deployment of AI in sensitive areas like mental health and societal integrity.

In a 2024 study, Castillo-Campos and her team analyzed various generative AI tools, finding that these systems often generate rather than merely extract words from input data, producing subjective or opinionated terms. Approximately 7% of the words generated by these AI tools displayed subjective bias (Castillo-Campos et al. 2024). This underscores the need for transparency in training data and human oversight to ensure accuracy, ethical standards, and bias mitigation in generative AI systems, even as they operate at unprecedented speeds and scales. In AI-driven media, these biases could shape the narratives consumed by large audiences, influencing perceptions and reinforcing problematic stereotypes.

As Floridi emphasizes (2023), AI's crucial limitation is its lack of real reasoning. Large language models operate primarily on syntactic rules rather than understanding semantic content, which means they simulate intelligence without actual reasoning. This detachment between a technology-driven agency and the concept of intelligence (Vilaca et al. 2024) highlights the boundaries of generative AI's capabilities. It raises significant questions about how these systems are applied in contexts requiring genuine comprehension.

Thimm et al. (2024) examined how AI systems that mimic human-like communication can evoke matching or conflicting emotional responses in users. This anthropomorphization blurs the line between human and artificial emotional intelligence, complicating how users perceive and react to AI-generated content. Generative AI models, especially those designed to engage emotionally with users, must handle these complexities carefully to avoid creating misleading emotional experiences.

Moreover, AI systems, particularly those using reinforcement learning, rely on a *reward system* to optimize their decision-making. These systems operate in uncertain environments where outcomes are unpredictable. AI agents receive rewards based on the success of their actions, which encourages learning through trial and error. This feedback-driven model optimizes content creation based on engagement metrics like likes, shares, and user retention. By receiving rewards for successful content generation, the AI adapts its outputs. Learning from this human impact, generated and *synthetic emotions* drive the following feedback loop without genuine feelings.

The potential for *emotional dependency* on AI is a psychological condition where an individual's emotional well-being is heavily reliant on the technology's support, attention, or approval, as examined in the study on the chatbot Replika, designed for emotional companionship (Laestadius et al. 2022). The intense emotional connections some users form with AI entities mirror those found in human relationships, with significant mental health implications. This emotional entanglement with AI raises critical concerns about the psychological impact of such technologies, particularly for vulnerable individuals seeking solace in synthetic companionship. Likewise, social bots can disproportionately amplify low-credibility content, manipulating human vulnerabilities and influencing societal narratives (Shao et al. 2018). This manipulation challenges the integrity of information ecosystems and highlights AI's potent ability to sway emotions and beliefs on a grand scale.

"Does AI have a *dirty mind*, too?" – asks Cichy (2019), probing AI systems' emotional boundaries and interpretative capabilities. Cichy's concept highlights how AI can intentionally generate pornographic or aggressive content based on input data. This raises significant concerns about the psychological impact on users. AI systems that produce inappropriate material risk exacerbating issues like desensitization to violence or objectification, further complicating the relationship between human emotions and synthetic interactions and challenging the reliable design of AI technologies.

Predicting Futures

The future of AI isn't merely about advancing technology—it's about transforming the world to better align with AI's capabilities. Predictive analytics directly influence this alignment through both human and synthetic inputs. This approach suggests that stability-based systems could enhance AI's performance and foster more consistent, reliable outcomes across different sectors. Accordingly, David Lewis (2023) suggests a "stable world" concept that enables AI's effectiveness in a predictable, structured environment. This principle provokes reconsidering how we design societal systems, calling for more control and predictability to harness AI's broad potential.

As generative AI constructs a collective historical and real-time memory, it reflects and transforms societal narratives to this stable world and technology-driven predictions. However, these systems can reinforce preexisting beliefs, leading to predictable or non-predictable self-fulfilling prophecies. The most unpredictable outputs are the "black swan" events. These rare and unforeseen disruptions can be, for example, caused by GPT-created disinformation or deepfake videos of public figures with unintended profound consequences in the society or economy. Moreover, if AI-driven predictive algorithms begin to mirror the above-mentioned "memory effect," the resulting societal perceptions may grow increasingly unpredictable, significantly amplifying the likelihood of black swan events and unforeseen disruptions.

In line with this, Makridakis (2017) presents a considerable vision of the future with AI. One path leads to a *utopia* where AI augments human capabilities and addresses global challenges. The other veers toward a *dystopia* marked by job losses and privacy erosion in surveillance states. This twofold scenario encapsulates the inherent unpredictability of AI's impact, underscoring the need for cautious optimism and strategic planning.

Accordingly, Danaher's (2022) concept of "modest techno-optimism" advocates for a balanced perspective on technology's role in shaping the future. It acknowledges the critiques of unchecked techno-optimism while defending the potential of technology, including AI, to enhance human well-being. This nuanced stance encourages a pluralistic approach, integrating diverse viewpoints and values to manage the complexities of technological advancement.

With a pragmatic approach, the Gartner Futures Lab (2023) adopts a comprehensive approach to future forecasting through the *TPESTRE Model*, which examines technological, political, economic, social, trust/ethical, regulatory, and environmental trends. This model offers a holistic framework for anticipating future developments, emphasizing the interconnectedness of various domains and their collective influence on the trajectory of an AI-driven society. Even if this strategic approach incurs additional costs, it is essential for tackling fast-evolving trends and building adaptability, resilience, and innovation. These are the key areas where consultancy firms see substantial business opportunities amid unpredictable AI-driven shifts.

Predicting the future in the context of AI and technological evolution involves grappling with a spectrum of possibilities, from transformative advancements to existential challenges. The discourse traverses various domains, from the pragmatic considerations of AI adoption to *stable world principles* or the *regime of hope* with essential principles, optimistic beliefs, and progressive potential. A blend of cautious optimism, strategic foresight, and inclusive dialogue should emerge to engage with these discourses. Without this, only the dangerous overoptimism or the islands of social responsibility remain.

The signs of the future often surface in current dynamics. In a symbolic sense, these signs are available via high-profile events marking upcoming issues, power structures, and looming disruptions. The AI Safety Summit in 2023 was this kind of noticeable high-profile event where prominent figures from global geopolitics and big tech met. These early summits signal emerging issues that demand significant attention for innovation strategy and economic stability. Their signals help to identify less uncertain future scenarios or surface wild card events, which can preempt chaotic and unpredictable black swan occurrences.

Societal Perception

The societal perception of generative AI and AI media is a multifaceted phenomenon that encapsulates this transformative technology's hopes, fears, and controversial developments. Society stands at this constant crossroads, considering the technology's potential to transform industries and everyday life while disrupting privacy, employment, and valuable human interaction.

Research on public perceptions often emphasizes the role of mass media in constructing societal narratives and influencing interpretations of technologies. For instance, one research scope examines whether the *media's portrayal* of technologies is driven by hype (Vicsek et al. 2025). Another investigates workforce and policy challenges through social media discussions (Li et al. 2024). Likewise, optimist and skeptical narratives also appear. News outlets, for instance, highlight extreme promises and pitfalls, contributing to a complex public discourse. However, this portrayal can lead to inflated expectations or undue skepticism. Another example is optimism surrounding AI's potential to enhance efficiency and alleviate workloads, as observed in Ashworth's (2023) survey. This optimism, however, is uniform across only some sectors of society.

In parallel, a significant knowledge gap exists in the professional perception of transformative technologies, where many still need to discover challenges and possibilities. Generative AI and AI media are critical sectors; thus, tons of online upskilling courses have found their business potential in this niche. However, companies with these upskilling profiles must also update these courses frequently, following the demand for new human skills with even more complex AI models. This gap underscores a broader challenge: the need for widespread AI literacy and education to demystify the technology and empower individuals to harness its potential responsibly.

The emergence of deepfakes is a vital part of societal self-perception. This technology directly mirrors the darker side of AI's capabilities for creating alternative realities and memories. Deepfake technology starkly illustrates how AI-driven or synthetic content can erode trust, setting off alarms about authenticity in AI-driven or synthetic content. Christopher (2023) suggests that trust issues of AI-generated content must be critically studied for future-oriented solutions. He highlights the case of an Indian politician accused of corruption through AI-generated audio clips. He underscores with this example a need for robust verification mechanisms in societal perception. However, this challenge extends beyond politics, touching on every aspect of generated and mediatized AI mechanisms and raising several critical questions about the consequences. Recalling Lewis's concept about the stable world from this chapter (2023), predictable changes and practical concerns shape the perception of technologies. Confirming this principle, more profound perception studies will be crucial to understanding media messages and their verification.

Popular narratives around AI, as explored by Wang et al. (2023), also play a pivotal role in shaping public perceptions and are deeply influenced by cultural, socio-economic, and political contexts. These narratives, ranging from dystopian fears of job displacement and surveillance to utopian visions of AI solving global challenges, reflect societies' diverse hopes and concerns toward AI. The variation in these narratives across different countries highlights the importance of context, underscoring the need for a nuanced approach that considers each society's unique priorities and challenges.

The anthropomorphic framing of generative AI tools also significantly shapes societal perception. This framing attributes human-like traits and emotions to GenAI systems, influencing public perception, trust, and ethical considerations regarding their capabilities and responsibilities. Research by van Es and Nguyen (2024) reveals, for example, that ChatGPT is commonly depicted as a "knowledgeable and helpful assistant," cultivating trust and a sense of reliability in human-AI interactions. However, such portrayals risk inflating expectations, leading users to overestimate AI's capabilities and then see it as an infallible, "friendly assistant." Recalling the concept of dangerous optimism from the chapter "AI media," individual perceptions mirror societal attitudes, potentially embedding AI authority and influence (Feher et al. 2024).

Impact Project

The MIT Climate & Sustainability Consortium (2024) initiative, "Considering the Environmental Impacts of Generative AI," addresses the sustainability issues caused by the rapid expansion of generative AI. The project aims to promote responsible development in the face of these challenges. With GenAI systems transforming media content creation and distribution, the project emphasizes reconciling AI's potential for creativity and productivity with its high environmental costs. The initiative advocates for a progressive and sustainable approach to AI development, ensuring that these advancements do not contribute to environmental degradation. Moreover, the long-term goal is net-zero emission to balance the amount of greenhouse gases produced and the amount removed from the atmosphere.

Generative AI models, particularly those underpinning generative systems, are incredibly energy-intensive, contributing to increased electricity consumption and rising carbon footprints. This is particularly alarming in media, where AI-driven content production—spanning from news generation to entertainment—relies on vast datasets and resource-heavy algorithms. The reliance on such resource-intensive processes and unsustainable consumption patterns exacerbates environmental pressure. The MIT initiative addresses these critical long-term challenges by proposing a balanced framework that assesses GenAI's benefits while accounting for its societal and ecological costs.

To mitigate these environmental impacts, the MIT Climate & Sustainability Consortium has introduced a benefit-cost evaluation framework focused on reducing the energy consumption of AI models without sacrificing operational efficiency. The initiative advocates for adopting renewable energy sources to power AI-driven media processes and encourages the development of more energy-efficient computational models. Additionally, it calls for enhanced transparency in supply chain management to address the macroeconomic impacts of GenAI's rapid growth. Through this structured benefit-cost evaluation, the initiative promotes a balanced approach that leverages AI's productive potential while ensuring it aligns with sustainability goals. This forward-thinking strategy underscores the need for cross-industry collaboration to implement scalable solutions.

This initiative holds the potential to profoundly impact how generative AI is developed and deployed. By integrating sustainability as a central consideration, the project encourages industries to reconsider their reliance on resource-intensive technologies. It also seeks to shape future policies and ethical standards, advocating for responsible AI use across various sectors. Accordingly, the initiative encourages global standards for sustainable AI innovation, ensuring that AI-driven society evolves in a sustainable, inclusive direction.

This initiative provides a vital roadmap for the future of generative AI, reducing environmental uncertainties. It highlights the need to balance AI's benefits and costs, offering a scalable framework for responsible development.

Continuing stakeholder engagement and industry collaboration will be essential to ensure AI technologies support a sustainable and inclusive future where innovation is aligned with environmental and societal well-being.

Reflective Art

Uncertainties and sustainability concepts are well-represented in tech-driven art to touch the cognitive-intuitive level of the above-discussed complexities. By exploring these themes, artworks not only engage with the immediate effects of technology on society but also provoke deeper contemplation on long-term implications, urging a more conscious approach to technological development and its integration into the fabric of human life. This intersection of art and technology fosters a critical dialogue essential for understanding the complex domain of modern innovations and their repercussions on our collective future.

Bio Morphosis by Stanislas Chaillou (2024) explores the fusion of various plant species through AI, presenting a visually stunning yet ethically complex exploration of biotechnological possibilities. It raises questions about the limits of artificial manipulation of natural forms, highlighting the uncertainty in tampering with biological diversity and the broader implications for ecological systems and biodiversity conservation.

The *AI-Generated Magazine Cover for Cosmopolitan* project—Karen X. Cheng and OpenAI's DALL-E 2 collaboration—represents a significant moment in the intersection of AI and media (Liu 2022). It challenges traditional notions of creativity and authorship, pushing the boundaries of AI's role in content creation. This project raises uncertainties about the future of design professions and the authenticity of creative works, stirring debates on innovation perception and the evolving role of AI in artistic expression.

Victimization in the Metaverse by Deepfake (Stavola & Choi 2023) addresses the deepfake phenomenon within the burgeoning metaverse, underscoring the pressing concerns around identity theft, privacy violations, and the spread of disinformation. The project highlights the technological challenges in detecting and regulating deepfakes. It emphasizes the psychological toll on victims, pointing to the urgent need for comprehensive strategies to manage these risks and safeguard individuals in virtual spaces.

Collectively, these reflective projects present the complexities and uncertainties inherent in the intersection of media and generative AI. Ecological, societal, and individual uncertainties reflect on the boundaries of technological intervention in nature, where humans are also a part of nature and technology. Exploring the intersection of AI and creativity, all three projects raise questions about the ethical implications and uncertainties of AI's increasing role in media, biology, and identity, even its impact on societal trust and innovation opportunities. While "Bio Morphosis" focuses on ecological and biological manipulation, "Cosmopolitan" challenges creativity and authorship, and "Victimization by Deepfake" highlights identity theft and psychological harm in synthetic spaces.

These diverse artwork projects still highlight emerging uncertainty issues, enabling the interpretation of generative AI's impacts on human creativity, identity, and ecology. These artistic approaches engage diverse audiences and foster a deeper public understanding of generative AI-media-climate literacy.

Conclusion

Uncertainty, in its essence, intertwines intricately with the fabric of our evolving society, particularly as we explore the advancing frontiers of technology. The dual nature of AI innovation emerges as a double-edged sword, presenting groundbreaking opportunities and formidable challenges.

The innovation is marked by complex narratives, where the beauty of uncertainties and the potential of neuromorphic computing collide with the vulnerabilities inherent in reward systems. This dichotomy underscores the importance of metareasoning in steering AI's capabilities toward beneficial outcomes while being mindful of the ethical deliberations accompanying such advancements. Content mining for advanced model training, synthetic media products, and memory effects blur the lines between reality and fabrication, further complicating our interaction with technology and highlighting the need for a *stable world principle* to address *elusive possibilities* and a predictable future. Social and environmental sustainability is a central consideration in this discourse, emphasizing the need for a balanced approach that deeply understands the planet's ecological boundaries and the integrity of the societal fabric.

Predicting futures is inherently uncertain, yet the uncertainty ranges from the more controlled nature of predictive analytics to the unpredictability of black swan events. Scaling these factors introduces speculative risks and foundational shifts. Societal perception, media narratives, and reflective art significantly shape public acceptance, which then feeds back into generative AI or synthetic media, reinforcing these structures. This feedback loop between perception and AI-generated content creates an *ouroboros effect*, where technology perpetually interacts with and reshapes its foundational knowledge. At each stage, complexity deepens, raising questions about whether the system may eventually consume itself through recursive, synthetic insights—drifting further from traditional, fact-based human knowledge production. Hallucinations, simulated truths (Burke & Schellmann 2024), and confidently presented yet unfounded "facts" merge into this ouroboros effect, displacing authentic or original information with AI-generated narratives. This loop mirrors how biases, gossip, propaganda, and media manipulation have historically shaped public opinion. However, generative AI accelerates and globalizes these processes, potentially leading to polarized viewpoints or power shifts. The essential question, thus, remains: How can we govern and manage this convergence of human and machine-driven uncertainties, particularly in high-stakes domains?

Scholarly Debates

For Cultural and Social Sciences

- How do popular narratives around generative AI influence public perceptions and societal expectations of technological advancements?
- How does the concept of memory effect connect to generative art and media?
- What are the generative AI and AI media uncertainties that challenge societal-environmental sustainability?
- Why is emotion AI one of the most controversial topics in societal and individual contexts?

For Science, Technology, and Engineering

- How can scientists and engineers effectively manage the uncertainties inherent in generative AI development?
- What are the main challenges in ensuring AI systems promote social and environmental sustainability?
- How can neuromorphic computing introduce further uncertainties for human cognition?
- Why do measurable probabilities and elusive possibilities generate complicated distinctions between reality and fabrication?

For Business, Policy, and Innovation

- How can policymakers address the uncertainties of AI media's impact on societal stability?
- What policy frameworks are needed to manage the risks and uncertainties associated with generative AI?
- What business models mitigate uncertainties in generative AI markets?
- What are the long-term implications of uncertainties in AI media for business and the economy?

Strategic Takeaways

For Academia

- Societal-Environmental Studies: Examine ecological systems and related socio-political frameworks to address generative AI's uncertainties holistically.
- Predicting the Unforeseen: Investigate uncertainty reports of AI-driven media's societal and cultural impacts.
- The Economy of Synthetic Media: Analyze how emotion AI transforms the media industry, labor markets, and global economies.
- Generative Narratives: Compare popular media's portrayal of generative and synthetic technologies.

For Industry

- Improving Decision-Making: Implement metareasoning in generative AI to improve autonomous decision-making in ambiguous situations.
- Assessment of Models: Assess how AI media trends influence business models.
- Balancing Sustainability Aspects: Explore strategies for how the environmental impact can be reduced using generative AI.
- Mitigating Disinformation Risks: Develop robust verification mechanisms to combat disinformation risks.

For Policymaking

- Frameworks for Control: Develop a generative AI policy for assessing societal uncertainties.
- Global Uncertainty Management: Foster international cooperation to manage uncertainties of AI media impacts.
- Sustainability Balance: Implement responsible business and public policies for environmental sustainability of generative AI.
- Public Discourse: Engage public discourses on synthetic media communication of diverse societal values and interests.

5

ETHICS

Defining Ethics

Generative AI and AI media ethics focus on the responsible and explainable creation and distribution of synthetic outputs. These ethics balance innovation with risks such as harm, misuse, and manipulation while ensuring fairness, transparency, and inclusivity to safeguard societal values. The related ethical considerations also address the dual-use dilemma, disinformation, bias, and public trust and protection of privacy. Central to these frameworks is agent behavior within a social contract, highlighting the importance of human-centered design and integrative ethics. Artworks in this context primarily explore generative AI's influence on identity and historical truth, emphasizing the need for critical reflection.

Introduction

As generative AI transforms the way in which content is created, disseminated, and consumed, it also radically changes how information can be accessed (Natale & Cooke 2021). Generative morality and human values are central to related societal debates and technological advancement—from ethical AI strategy to human rights. This is particularly true if emerging generative AI services prioritize user-friendly outputs at the expense of accuracy, data privacy, or social-societal well-being.

Consequently, this chapter aims to explore the principles and challenges of trusted generative AI and AI-driven media, probing into the complexities of ensuring that the technology operates within an ethical framework. The focus is on how generative AI ethics can be codified for the long term while the rapid evolution of the technology is evident. The question is of how *responsibility*

DOI: 10.4324/9781003591023-6

and *explainability* can become more a foundational practice in development and not mere buzzwords. This approach touches on generative AI's harms and misuse, setting off alarms about biases and inequalities (Hoffman et al. 2022).

The analysis highlights the moral challenges of fluid, context-based and relational technology, urging a shift toward more ethical practices in generative AI development and AI-driven content production. This involves a collective effort for ethical AI and ethics by design in emerging technologies. The goal is to address the intricacies of constant and emerging concepts in AI ethics, contrasting different approaches. This reflection comprehensively summarizes the leading or emerging concepts and proposes an integrative ethical framework.

Generative Morality and AI Ethics

Generative morality uses ethical principles and moral frameworks as inputs, turning them into prompt-driven outputs—incorporating societal values into generative AI through dynamic interaction and creative processes. According to Sibert (2020) of TU Delft, morality should be embedded in AI systems in order to align with traditional ethical theories, and thus prevent potential abuses and bring technological outputs into line with societal values. A blend of utilitarianism and Kantian deontology could underpin this approach, particularly in the context of autonomous AI systems.

Utilitarianism and Kantianism, accordingly, are two inputs of generative morality that consider the way in which individuals and institutions feed machines with their blended data. *Utilitarianism* emphasizes the maximization of overall societal benefits, potentially supporting AI-generated media and the content industry in improving user or audience engagement. The benefits could outweigh the harms, even when there is a risk of spreading mis/disinformation. Conversely, *Kantianism* prioritizes individual rights and duties (Mittelstadt 2019), insisting that generative systems respect autonomy and avoid manipulation regardless of societal impact. In other words, AI content that deceives or exploits users would violate Kantian ethics, which emphasize transparency and truth. A balanced combination of utilitarianism and Kantianism, therefore, would result in a normative ethical framework that maximizes societal benefits and protects human dignity, privacy, and truth. Such a balance is essential for shaping generative morality in society.

Illustrating the issues, generative models like Bhagavad Gita GPT or BibleGPT attempt to interpret complex transcendent values but often oversimplify cultural and ethical nuances due to limitations in pattern recognition. Sanjeev (2023a) suggests that *religious AI*, while promising, requires human oversight to prevent rigid interpretations of spiritual values and ethics. Another example is GenAI-supported behavioral and mental health, where the technology faces profound challenges due to the divergence between societal and individual values. The question is of how generative AI can interpret different mental health needs while standardized care models are also required. Sezgin and McKay (2024) highlight that while GenAI can support personalized therapeutic interventions, it often struggles with balancing broad societal norms against individual emotional and behavioral needs, complicating its role in customized care.

In addressing AI's socio-ethical challenges, the dominant focus in leading scientific concepts is the operationalization of *democratic values*, principles such as freedom of speech, equality, justice, and civic participation. As Reisach (2021) suggests, ethical principles must be entrenched within AI systems to safeguard democratic societies, particularly in the context of political manipulation via social media. Utopia (2023) offers a blueprint for an ethical AI framework, highlighting humans as the ultimate arbiters of AI's *moral compass* and for the ethical principles guiding AI systems to align with human values. This framework emphasizes shared responsibility, honesty, and equality as foundational pillars, ensuring that AI respects societal standards while promoting trust and transparency.

Coeckelbergh (2023) prompts a shift from the traditionally static view, arguing for a performative perspective in which moral status is not an intrinsic attribute but rather an outcome shaped by our linguistic and performative acts. This conceptualization acknowledges the fluidity of ethics in the age of generative technologies. Moreover, it recognizes that moral frameworks, including human rights, are not fixed entities but living constructs that require continuous engagement and reaffirmation by society. This approach, if it is integrative and not just dynamic, is helpful in the construction of ethics for generative technologies.

As Kotliar and Rivnai-Bahir (2023) argue, in terms of *business-driven morality*, data-driven corporate goals and profit motives challenge traditional ethical frameworks. This raises concerns about the way that profit-driven considerations in data science shape the design, implementation, and regulation of AI systems, often prioritizing financial gain over broader societal values. The influence of business-driven morality in AI ethics can shift public perceptions of fairness and accountability, potentially undermining trust in technology-driven industries. This approach is complemented by the work of Fabbri (2023), who advocates for *pro-ethical design* in recommender systems, meaning the intentional development of AI systems that prioritize ethical values such as fairness, transparency, and societal well-being over profit-driven motives. This approach challenges business-driven morality, where corporate interests often take precedence, by advocating for AI that aligns with broader societal needs.

In this way, *generative AI ethics* form the interplay between business-driven morality and pro-ethical AI design in the *society-human-machine interaction*. This interplay can be balanced if business-driven morality for profit maximization (Kotliar & Rivnai-Bahir 2023) and pro-ethical design for fairness (Fabbri 2023) together support the reflected cultural diversity and deeper societal values without an oversimplified ethical construction (Qadri et al. 2023, Nelson 2023). Accordingly, the tension between generative morality and financial gain drives a critical need for transformative deliberation (Bender 2024).

Dual Use Dilemma

As examined by Koplin (2023), AI-generated content has become a transformative force in content creation by using AI algorithms to produce synthetic outputs. This approach highlights the *dual-use dilemma*, historically linked to nuclear and biological technologies. On the one hand, *AI generators* enable immense creativity and productivity; while on the other hand, they raise ethical concerns, such as the potential for misuse in spreading disinformation or harmful content.

In this context, AI's generative text serves as a potent reminder of technology's potential to uplift, undermine, inform, and deceive. The dual-use nature of AI text generators requires a robust ethical analysis that extends beyond the capabilities of the technology itself. It questions the balance between the beneficial uses of AI language models in constructive applications, such as text summarization and creative assistance, against the backdrop of risks associated with malevolent uses, including the propagation of fake news and disinformation campaigns. As highlighted by the Institute for Ethics in AI Oxford (2023), the dynamic interplay of these uses illustrates the acute need for an ethical compass for every societal actor to understand generative and interactive AI.

The fundamental dilemma is one of how AI can be ethically used and potentially misused. Drawing parallels with *the trolley problem*—a classical ethical puzzle discussed by Narayanan (2018)—illustrates the difficulties faced in AI development. The choice to divert a runaway trolley toward one person to save five others echoes developers' decisions: whether to advance a technology with the knowledge that it could be repurposed for harmful ends.

This dilemma is not a new puzzle; just consider nuclear physics or genetic engineering. AI is simply the latest entrant into the arena of dual-use technology, carrying with it the potential for both extraordinary benefits and unprecedented risks. Karnouskos (2022), for example, notes that AI can augment human capabilities and solve complex, context-specific problems, yet it raises concerns about robot-to-robot interactions and their implications for law and society. However, this dilemma is intensified by the fluid nature of generative AI, where learning models adapt in real time based on new inputs, making it a moving target. GPT systems, for instance, have ever newer version numbers evolving continuously, altering outputs as they learn from data streams. This makes it difficult to precisely calculate the dual-use risks of AI, which can only be estimated, increasing uncertainties. As discussed in the last chapter, these complexities demand careful attention in AI governance and ethical frameworks.

The fluid nature of generative AI calls our attention to its continuous adaptability at the societal and individual levels. Learning models evolve in real time based on new inputs, making them a moving target. This dynamic process allows generative AI to generate diverse outputs and respond flexibly to changing data, societal trends, and individual interactions. As a result, it constantly transforms traditional content creation, decision-making, and communication processes, challenging static ethical frameworks and making regulation more complex.

The fluid nature of generative AI results in *fluid ethics*, where evolving AI-generated outputs challenge static ethical frameworks, making regulation and oversight increasingly complex. This can be broken down into pro-ethical designs in well-defined, coherent ethical systems on a context-specified basis. These systems are networked and interconnected rather than structured hierarchically, allowing for a more flexible and adaptive approach to emerging ethical challenges. *Fluid GenAI ethics* describes an adaptive, context-specified framework where ethical principles and guidelines shift in real time based on new data inputs and societal changes, which makes it a moving target for regulation. This concept aligns with the experience of the continual evolution of learning models (Koplin 2023), which renders fixed ethical systems insufficient. By integrating pro-ethical design into a networked structure rather than a hierarchical one, fluid generative AI ethics allows for more responsive, well-defined, and interconnected approaches to address the ethical complexities posed by ever-evolving AI technologies (Nelson 2023). Although the dual-use dilemma remains, uncertainties can be reduced by fluid GenAI ethics.

Trusted GenAI Principles

The principles of transparency, fairness, inclusivity, accountability, reliability, responsibility, and explainability serve as essential guideposts as we learn even more complex AI systems. These principles guide socio-cultural construction throughout the AI lifecycle, ensuring ethical alignment by addressing bias, sustainability, and governance within organizations while prioritizing societal impacts. As Baxter highlighted, these principles are foundational in managing the complex risks associated with generative AI, ensuring that AI systems are developed and utilized to uphold ethical standards and societal values (Baxter & Schlesinger 2023).

The cornerstone principle of transparency in generative AI demands clarity in the operation and decision-making processes of AI systems. This transparency is crucial in demystifying AI technologies and creating an environment in which users can trust and understand the mechanisms behind AI outputs, after promoting inputs. The cornerstone principle of transparency addresses challenges posed by the Trust Paradox (see chapter "Transformation") while promoting frameworks that improve user understanding and trust in AI outputs.

The call for transparency is echoed in the Stanford AI Index (2023), which underscores the increasing attention paid to ethical considerations in AI research, emphasizing the need for AI systems to be open and understandable so that fluid GenAI dynamics do not exacerbate societal inequities as these systems rapidly adapt and evolve, further complicating ethical oversight.

Fairness and inclusivity together ensure equitable participation in innovation processes if they extend beyond technical accuracy or mere societal ideals.

These principles are about ensuring that AI development reflects the diverse needs of all societal groups, not just the privileged few. Narayanan (2018) argues that fairness in AI must consider the nuanced impacts on various communities, pushing developers to incorporate diverse perspectives. Without inclusivity, AI risks perpetuating bias, reinforcing discrimination, and distorting representation, particularly in generative AI and synthetic media, where biased data can rapidly propagate harmful stereotypes or content in a fluid way.

Accountability is also crucial in the generative AI ecosystem, ensuring developers and operators are responsible for the unpredictable outputs these systems generate. Chhillar and Aguilera (2022) highlight the unique governance challenges of generative AI. They call for frameworks that align with ethical standards, mitigate risks like mis/disinformation and bias, and meet societal expectations to maintain public trust.

Reliability, as discussed by Appelganc et al. (2022), is fundamental to building trust in AI systems. This principle underscores the importance of AI technologies operating consistently and reliably, especially in critical applications, where failures can have significant consequences. In generative AI, reliability is essential because these systems generate autonomous and synthetic outputs. Wu et al. (2022) illustrate in their identity research, for instance, how human-AI interactions and contextual factors shape trust. Their findings emphasize that consistent AI performance across various applications is key to cultivating trust. This trust reinforces the need for reliable and transparent AI systems in diverse settings.

The principle of responsibility is intricately linked to the ethical deployment of AI technologies. Ashrafian (2015) proposes a forward-thinking perspective on AI and robot responsibilities, suggesting the need for AI systems to adhere to ethical frameworks that prevent harm and promote the well-being of all beings. This principle is essential in ensuring that fluid GenAI dynamics address issues such as mis/disinformation, unintended surveillance impact, or intellectual property problems, which as these systems continuously evolve present new and complex challenges for ethical governance.

Explainable AI is designed to make the outputs and decisions of AI models understandable to humans. It boosts trust, transparency, and debugging in AI systems by clarifying the way in which decisions are made (Appelganc et al. 2022). Generative Explainable AI, as discussed by Schneider (2024), adds complexity due to the autonomous, creative, and synthetic outputs of generative models, producing hallucinations and requiring output verification or specific GenAI ethical standards.

These principles are essential to building trust as we increasingly rely on AI systems. These principles, as also discussed by Fjeld et al. (2020) and Gartner Futures Lab (2023), guide AI development to ensure ethical alignment with societal norms and values. In generative AI, these principles are fundamental but come with unique challenges. The less transparent inputs and unpredictable outputs of GenAI make maintaining accountability and explainability more complex, as Schneider (2024) highlighted.

How can this complexity be handled to achieve societal trust of GenAI? The efforts of *generative AI brokers* could be relevant in response, acting as they do as intermediaries between developers, policymakers, and the public, ensuring that generative AI systems meet ethical standards. These brokers oversee the deployment of autonomous outputs, addressing risks like bias and mis/disinformation. Brokers balance transparency, accountability, and societal impacts while adapting governance frameworks to manage evolving challenges such as unintended consequences in content generation and privacy concerns, and ensuring responsible use in dynamic AI environments. However, we need many more GenAI brokers. A key challenge is equipping them with the knowledge to critically analyze AI developments while maintaining a highly skeptical approach to match the emerging challenges of generative technology.

Harm and Misuse

The deployment of generative AI without stringent ethical guidelines poses the risk of actual harm and reflects a troubling trend of misuse that capitalizes on the vulnerabilities of tech societies. In this context, various interests of sociocultural diversities, business-driven ethics or incompatible moral senses challenge society.

Baxter (Baxter & Schlesinger 2023) sounded the alarm on the unintended consequences of generative AI, a sentiment that resonates throughout the annals of disinformation history, from the deceitful campaigns of ancient Rome to the "Great Moon Hoax," culminating in the contemporary issues surrounding AI chatbots (Posetti & Matthews 2018).

The pervasiveness of AI brings to light the critical themes of ethics and business pressure, or rather, the lack thereof. Christodoulou and Iordanou (2021) discuss the inertia of AI developers and producers in addressing ethical issues, an inertia that is often masked by *ethics washing*. This practice represents superficial adoption of ethical frameworks or guidelines, creating a false image of responsibility without genuinely addressing deeper societal issues. In the shadows of these ethical quandaries, moral workers toil, as evidenced in Poland's *fake news factories*, where the human cost of data filtering is measured in mere dollars (Stahl et al. 2023, Feher & Kiraly 2017).

Compounding the ethical issues are instances of synthetic content amplifying *societal inequities*, from the sexualization of women in avatar-generating apps (Kruger 2023) to biases in generative content creation in Large Language Models (LLM). These are theoretical concerns and real-world manifestations of AI misuse that discriminate, invade privacy, and disrupt the sanctity of human dignity (Makridakis 2017, Stahl et al. 2023).

Sison et al. (2023) underline the potential of ChatGPT as *a weapon of mass deception*, or *manipulation*, echoing the need for a *human-centered AI* framework that prioritizes ethical guardrails to prevent deceptive applications. This is where the interdisciplinary approach advocated by Yu et al. (2023) becomes crucial; they call for a unified framework to address mis/disinformation and correction sharing. This framework must be conscious of the differential impacts of AI, such as the nuanced complexities of *helicopter research* (Stahl et al. 2023), where external researchers exploit local data without considering meaningful collaboration or whether it benefits local communities. A positive counter-example in generative AI is a proactive approach in moderating harmful content, implementing rigorous filtering and prompt restrictions to prevent disinformation. This includes developing preemptive strategies such as safety mitigations and ethical reviews to ensure outputs remain aligned with societal values.

The ethical challenges, as Christodoulou and Iordanou (2021) argue, are not confined to privacy and discrimination but extend to the effects of nationalism and populism. This populist rise threatens to destabilize the democratic fabric, demanding a reevaluation of our ethical frameworks within AI and big data—frameworks that must address the diverse and often conflicting ethical perspectives across different cultures (Fjeld et al. 2020).

In addressing these diverse ethical challenges, one must consider the moral value of *privacy*, akin to property rights, where the individual's control over personal spaces and information is paramount, as Moore (2017) posits. Within this context, we must deeply understand the ethics of AI—not as a series of isolated dilemmas but as interconnected challenges that reflect the complex interplay between technology, privacy, dignity, and the collective responsibility to foster trust (KPMG & University of Queensland 2023).

It's no coincidence that high-level discussions around generative AI ethics have increasingly become central at human rights conferences, or that research centers of AI ethics encourage human rights symposiums. This versatile focus on human rights highlights generative AI's potential for harm and its potential impact on democracy. This trend underlines the growing recognition of societal risks posed by generative AI.

Agent Behavior and the Social Contract

An *AI agent* is a system or entity that interacts with its environment autonomously, perceiving inputs and making decisions to achieve predefined goals. The AI agent's behavior refers to autonomously processing information, learning from experiences, and making decisions based on its programming and objectives. As AI agents become more advanced, their behavior adapts to real-world contexts, making their actions less predictable. This progression raises essential ethical considerations, influencing societal systems and decision-making processes (Wooldridge & Jennings 1995), also in generative ways. To align with ethical standards, generative AI must operate within frameworks that guarantee responsible decision-making, ensuring that its outputs adhere to societal norms and values, and foster trust and transparency in its deployment and interactions with *human agents*. Accordingly, the critical ethical challenge lies in ensuring that AI agent behavior aligns with broader human values, particularly in contexts like the media, where generative agents can impact emotional engagement, content creation, and public discourse.

Relational ethics, as advanced by Coeckelbergh (2020), emphasize the contextual and dynamic nature of human interactions. In this ethical view, AI systems or agents are not isolated entities but operate within complex social and cultural frameworks. For instance, AI-generated media content can shape public perceptions and pose risks of disinformation or manipulation. Understanding AI agent behavior through a relational lens means evaluating AI's technical performance and its broader impact on societal, cultural, and psychological dynamics, including the way it influences decision-making and public trust in media. This holistic approach ensures ethical alignment across various contexts of AI deployment.

The concept of the *social contract* can help understand AI agent behavior in generative AI and synthetic media. Rooted in long-established theories (Rousseau 1762; Hobbes 1651), this contract can frame AI systems, especially those generating media content, as societal tools bound by ethical agreements between developers, users, and regulators. The importance of such accords is obvious, considering the way that generative AI is capable of producing vast amounts of synthetic content that significantly impacts public opinion and human behavior.

Accordingly, Rahwan's (2018) *society-in-the-loop* model proposes a feedback loop between society and AI developers to ensure these systems adhere to social values and promote transparency and accountability. This model becomes essential in media environments where generative AI can enhance democratic discourse or propagate disinformation. As the Boston Global Forum (2020) underscores, AI systems must respect human rights, privacy, and the rule of law—and mitigate any counter effects. Generative AI systems must align with these principles to foster a trustworthy ecosystem that mitigates the risks of bias, mis/disinformation, manipulation, and misunderstandings of the human-AI interaction process.

Considering this approach, relational ethics, while supporting logical decision-making in AI systems, faces challenges when applied to generative AI's nuanced, socially embedded behavior. AI systems must not merely follow rigid ethical rules but also adapt to dynamic societal expectations via prompted input-output systems. The convergence of relational ethics and the society-in-the-loop model suggests that AI systems must evolve based on societal feedback, integrating ethical and contextual concerns into their decision-making processes.

The social contract between human and AI agent thus emphasizes responsible innovation and balancing ethical standards with societal values. This approach ensures AI-generated content serves societal progress while minimizing harm and strengthening accountability across its lifecycle.

(Un)Bias and Generative Colonization

Biases swing like a pendulum between technological imperfection and mirroring societal shortcomings. Baxter's insight that 73% of professionals are concerned about biased outcomes from AI systems is the tip of the iceberg, hinting at a deeper, systemic malaise that extends far beyond the technical definitions of bias (Baxter & Schlesinger 2023). Kruger (2023) elaborates on this by revealing how even content moderation in generative AI, designed to be the bulwark against harmful content, can be circumvented, thus raising the question of what constitutes *normal* bias and who decides what is biased. In contrast, Kozyrkov (2023a, 2023b) offers a technical view by dissecting bias into statistical misalignment, sampling errors, and cognitive deviations. Yet, this compartmentalization belies the inherent messiness of bias in the real world. Here, bias is not a mere technicality; it is injustice personified (Narayan 2022). It emerges from the fabric of human societies, echoing prejudices and inequalities through the ostensibly neutral algorithms we create.

However, AI transforms algorithm-driven neutrality into a synthetic out-of-context reality, where inputs and outputs are substituted by statistically generated content. This shift creates further detachment from the original context, allowing biases to be amplified or distorted. Instead of reflecting societal biases, generative AI introduces new layers of complexity, where content is a prior bias and disconnected from the real-world nuances that traditionally ground decision-making processes. As a result, bias in generative AI is not just a technical problem; it is an ethical challenge that requires more than mere statistical corrections to address the broader societal consequences.

Synthetic content generated by AI often replaces original contexts, leading to colonization. *Generative colonization* is the replication and expansion of biased narratives in synthetic content, reinforcing existing societal power dynamics through AI outputs. The critical issue, in this case, is not merely what content AI generates but how these systems interplay with biased and unbiased contexts in an interaction with human inputs. Trained on biased datasets, generative AI often mirrors societal inequities, embedding and amplifying preexisting power dynamics. This leads to the spread of dominant narratives, supporting exclusionary perspectives. Platforms like Latimer aim to counteract this by amplifying historically marginalized voices (Davis 2023), but without addressing inherent biases, AI risks amplifying, rather than dismantling, these societal structures.

Decolonization (Iliadis et al. 2023) and addressing diversity issues (Davis 2023) become critical in this discourse. The bias detected in AI media (Aggarwal et al. 2020), the systemic nature of biased networks (Neuhauser et al. 2021), and the representation of underrepresented groups of generative content recall the risks of losing democratized knowledge and generating further imbalance in societies. The consequences are profound. Packin (2021) highlights the discriminatory potential of AI in social scoring, particularly against people with disabilities, while Feher (2022a, 2022b, 2022c) touches on the manipulation risks inherent in emotion AI. Kahneman (cited by Newel 2021) distinguishes bias and noise, suggesting that the unpredictability of human judgment, or *noise*, may be a more formidable adversary than bias itself. Chin (2021) reinforces this by arguing that reducing noise may be more effective in improving decision-making than directly combating cognitive biases.

Considering this bias-driven jungle of ethical questions, we stand at a crossroads. The challenge lies in mitigating the biases evident in AI systems and confronting the uncomfortable reflection they offer on our societal norms. And because there is no zero-bias system, the key question, then, is not whether a society can create unbiased AI but whether a society is willing to address the biases to use diversity meaningfully.

Responsible and Explainable GenAI

Responsible AI (RAI) and *explainable AI* (XAI) are crucial frameworks in generative AI, also for AI-driven media and content creation. RAI ensures that AI aligns with societal values like fairness and transparency, while XAI focuses on making AI's processes understandable and traceable (IEEE 2024; Schneider 2024). In generative AI models, these principles help address risks like bias, mis/disinformation, and content ownership. RAI maintains ethical accountability, while XAI emerges to offer clarity on AI decisions, making them vital in mitigating ethical risks in a rapidly evolving media ecosystem. AI-driven media also amplifies such concerns, as the speed and scale at which content is produced lead to significant ethical challenges. For instance, generative AI can streamline the creation of articles, videos, or images, but this automation risks mis/disinformation and a lack of accountability for the generated content. The importance of XAI in this context lies in making AI's decisions and processes transparent, ensuring that both consumers and creators understand how content is produced and curated. This is particularly relevant in an era of synthetic media, where AI-generated content may blur the lines between reality and fabrication, further complicating the foundation of trust in media.

Baxter highlights that while innovation in generative AI is essential for growth in the media and content sectors, constant oversight is necessary to align these advancements with ethical principles (Baxter & Schlesinger 2023). As generative AI models become increasingly integral to creative industries, the risk of deepfakes, biased outputs, and privacy violations grows. A proposition of a human-in-the-loop approach offers a viable solution, ensuring that generative models can be corrected, audited, and guided by ethical standards, minimizing harm while maximizing creativity.

Following RAI and XAI logics, a similar framework is proposed for generative AI. *GenRAI* (Generative Responsible AI) focuses on the ethical generation of synthetic content by ensuring that the AI systems involved in creating text, images, or media autonomously adhere to clear boundaries on mis/disinformation, intellectual property, distortion of context, and content manipulation. *GenXAI* (Explainable Generative AI) goes further by providing transparency in how these systems generate content, allowing users and stakeholders to trace how inputs are transformed into synthetic outputs and to ensure that biases, context distortion, or unintended misrepresentation are minimized (adapting Schneider's approach, 2024). These frameworks are particularly critical in generative AI, where large-scale, autonomous content production can exacerbate issues such as deepfakes, synthetic data misuse, and out-of-context media creation (Baxter & Schlesinger 2023). With the rapid pace of development in areas like gaming and content mining, the challenge is embedding these ethical considerations without stifling innovation, as many service providers prioritize speed and engagement over transparency and accountability (Iliadis et al. 2023, Davis 2023).

The principles of transparency and accountability must guide the operationalization of generative AI, requiring an interdisciplinary approach that bridges product management, data science, privacy, and legal frameworks. This comprehensive strategy aims to mitigate potential harms while amplifying the social benefits of AI, suggesting a move beyond the immediate scope of technology into normative ethics. How these principles adapt to specific technological contexts and whether universal ethical norms can be applied across diverse scenarios remains a critical question for technologists and ethicists alike. In line with this, Baxter advocates for constant oversight and automation of review processes in AI systems, emphasizing the need for human involvement in ensuring safety, honesty, empowerment, and sustainability (Baxter & Schlesinger 2023). To apply GenRAI and GenXAI, a generative AI value chain can also be developed as processes involved in training, deployment, and monetization of the technology, from data collection and model development to content creation, customization, and feedback-driven improvement. This comprehensive process and the human-in-the-loop approach is about refining AI's performance metrics and ensuring the technology aligns with human values and societal well-being.

The challenge of fortifying truth in generative AI and deepfakes, as addressed by the Harvard Carr Center for Human Rights Policy (2023), is emblematic of the need for robust tactics, strategies, and infrastructure in policymaking. This includes preserving and moderating content, scrutinizing images and videos, and ensuring equitable access to detection tools. Democratization of AI, thus, urgently requires international standards, especially in AI media (Keenan 2023), with data ethics frameworks or holistic data management (OECD 2020). Democratization and accessibility in preventing the monopolization of AI technologies is a parallel requirement (Jansen 2022). These perspectives are crucial in ensuring that AI aligns with human values, societal norms, and responsible regulation.

The *Huawei Moon Mode controversy* discussed by Zhang (2022) exemplifies AI's ontological and ethical implications in everyday technologies. It prompts us to reevaluate the nature of authenticity and representation in AI mediated systems, raising questions about the ethical intentions behind AI alterations and transparency. However, as Kerr et al. (2020) point out, the discourse on ethical AI is often performative, serving more to reassure stakeholders than to enact substantial change. This performative element in ethical AI discourse suggests a need for a more profound acknowledgment of AI's limitations and a commitment to genuine ethical governance and standards—also via International Organization for Standardization (ISO) management.

Integrative Ethics

Generative AI challenges the fabric of human-machine interaction, blending synthetic outputs with moral sense and our cultural and emotional realities. Managing AI ethics, therefore, is no longer about merely applying rules—it's about a radical negotiation of power, transparency, and societal values. GenAI isn't just code; it's shaping narratives, perceptions, and trust, making the ethical management of these systems a collective act of accountability. As Hirsch et al. (2020) emphasize, harmonizing the strategic benefits of generative AI with ethical responsibility demands an integrative approach in which diverse stakeholders continuously challenge innovation boundaries to prevent the erosion of public trust. Contextual transparency (Sloane et al. 2023) can also support the integrity of ethics if it enables accountability and informed engagement.

Companies are not just chasing compliance and competitiveness in the GenAI business; they have to sustain trust among customers and employees, preparing for a regulated future in which ethical guidelines are ideally values rather than constraints. This journey requires a tapestry of inputs, from hiring AI ethics officers to establishing ethics committees and conducting impact assessments. It is about constructing a bridge between lofty principles and the challenging realities of product development as professionals translate human rights into questions that can guide developers in crafting more ethical AI systems.

It is not just a moral imperative but a creative catalyst that ensures the human workforce is reflective of the well-identified human values that they serve. As Foley (2022) contends, this diversity is not just about culture or demographics; it includes the rich spectrum of human experiences, ensuring that AI systems serve justice as much as they serve functions. On the one hand, questions raised by Genevieve Bell in the TED Salon (2020) resonate as foundational pillars for ethical AI: autonomy, agency, assurance, interfaces, indicators, and intent. On the other hand, as the introduction to this book highlights, these are often just platitudes from the stage at prestigious conferences. Behind the scenes, competing interests, cost efficiency, and competitive advantages drive controversial deals. Despite media scandals exposing ethical lapses, the business-driven momentum persists. Several times the vision for GenRAI or GenXAI has been reduced to mere corporate social responsibility efforts or fine-tuning. This contrast between public messaging and backstage realities underscores the need to integrate ethics meaningfully and for the long term into (Gen)AI development.

Viewing ethics as an investment, as Bevilacqua et al. (2023) propose, reframes the narrative around AI ethics from a cost to a value-add, generating returns that span economic benefits, reputational fortitude, and enhanced capabilities.

In this light, ethical AI would not be a tax but an investment in the future. As an illustration, inclusion is an urgent moral imperative, particularly for marginalized demographics such as older adults aged 65 and above. Sourbati and Behrendt (2021) highlight the chasm between those engaged with AI technologies and those on the periphery. Bridging this AI divide is not just about technology access; it's about ensuring that all voices contribute to generative AI inputs. This is a once-only opportunity for marginalized groups and individuals to ensure their representation, harness the benefits of GenAI, and secure their visibility for future generations. Addressing the AI divide is crucial for ensuring equitable access and participation in the benefits of generative AI technologies for all segments of society.

AI media ethics also presents a narrative of transformation and challenge. As Trattner et al. (2022) note, AI has become a powerful actor in media production and consumption, shaping how information is crafted and shared. Ethical considerations in this domain are critical to maintaining an innovative and responsible media ecosystem. In AI media, control is also challenging. *Ethical scaling of moderation*, for example as in Udupa et al. (2023), is a blend of technology, ethnography, and community involvement to address the challenges of online hate speech. This approach demands an iterative, transparent, and equitable moderation process where the AI is a partner, not a panacea.

Considering the versatile approaches, *integrative ethics* is proposed here as a multifaceted framework to AI ethics, harmonizing top-down regulations with grassroots moral engagement. This framework encourages blending diverse ethical theories and practical strategies to address the complexity of AI systems that create synthetic content. Integrative ethics ensure that AI systems are aligned with societal values and emphasize fairness, transparency, and accountability while minimizing risks such as bias, mis/disinformation, and exploitation. Such a framework is vital in the rapidly evolving domain of generative AI, where traditional ethical boundaries often fail to keep pace with technological innovation (Feher 2023a) and dynamic input-output interactions.

Though global initiatives like UNESCO's consultations and the Rome Call for AI Ethics strive for ethical consensus in AI, they are often limited in addressing the full spectrum of cultural, economic, and social concerns (Chadwick et al. 2023). Integrative ethics, however, builds on this foundation by recognizing the need for a more fluid and adaptable ethical model. It incorporates insights from Rodgers and Nguyen (2022), who argue for the importance of multiple ethical pathways, allowing for a nuanced response to dilemmas. Moreover, Coeckelbergh (2020) and Sanjeev (2023c) emphasize the necessity of tackling AI's broader societal entanglements, including the risks of AI colonialism and data sovereignty, further supporting the integrative ethics approach.

By encouraging ethical flexibility and collaboration across stakeholders, integrative ethics fosters a future-oriented approach that ensures generative AI and AI-driven media are innovative and socially responsible, contributing to societal progress without compromising integrity. This integrative ethics framework demands significant investment, but the return is clear: Generative AI companies prioritizing ethical standards will dominate, gaining consumer trust, mitigating risks, and turning social responsibility into long-term profit and competitive advantage, in an ideal case.

Impact Project

The *Data School at Utrecht University* is at the forefront of addressing the critical ethical challenges posed by AI, particularly within the domains of public management and media (Franzke et al. 2021, Gerards et al. 2021, Siffels et al. 2022, Schaffer 2024). As AI technologies, including generative AI, continue to shape the way in which societies function, there is an urgent need for structured, forward-thinking approaches to ensure ethical deployment. This initiative focuses on developing and implementing responsible AI practices to address policy gaps and foster sustainable, ethical AI applications in the media and beyond.

As AI's influence grows, ethical concerns around generative AI have intensified, particularly in its application to the media and public discourse. Misinformation, disinformation, bias, and a lack of transparency in AI-generated content present significant societal risks. Current AI policies and frameworks are often insufficient to address the challenges of rapidly evolving generative AI models. These gaps in AI ethics and policy threaten to undermine democratic institutions and public trust in media. Without effective governance, the long-term impact of generative AI on public perception, privacy, and accountability could be detrimental.

The Data School's response to these challenges lies in its innovative tools like the *Fundamental Rights & Algorithms Impact Assessment* (FRAIA) and *Data Ethics Decision Aid* (DEDA), which provide structured, comprehensive evaluations of AI's societal impact. These tools are designed to assist policymakers and civil servants in making informed decisions about the ethical deployment of AI in media. FRAIA, for instance, ensures that AI projects adhere to fundamental human rights principles, while DEDA guides stakeholders in addressing ethical concerns at various stages of AI implementation. By integrating academic research with practical solutions, these frameworks promote transparency, accountability, and fairness in using AI, and generative models in the media in particular.

The Data School's tools have already gained traction in government organizations across Europe, setting a solid foundation for future AI ethics standards. Their long-term vision is to foster a global standard for ethical AI practices, particularly in generative AI applications within media. The initiative could redefine how AI technologies are governed, by aligning AI development with societal values and human rights, ultimately shaping sustainable and ethically responsible AI in media and beyond.

The Data School's work highlights the importance of integrating ethical frameworks into AI governance, especially in generative AI and media applications. In addressing technical and societal challenges, the initiative sets a precedent for future AI policies. Engaging with stakeholders at all levels—government, academia, and industry—will be essential in advancing this progressive approach and ensuring the ethical deployment of AI across sectors.

Reflective Art

Artwork in generative AI ethics is a powerful medium for illustration. It can embody abstract ethical dilemmas, giving a tangible form to concerns about perception and identity, often making the invisible algorithms perceptible and prompting public discourse. Each piece acts as a catalyst for reflection, opening up critical conversations about GenAI's role in shaping our collective future or altering our documented past.

Face Cages, a thought-provoking art piece by Zach Blas, presents stainless steel masks as a critique of biometric facial recognition's ability to cage identities within transformative parameters. It symbolizes the constraints and profiling marginalized communities face under the guise of technological advancement (Blas 2016).

In Event of Moon Disaster, by Francesca Panetta and Halsey Burgund of the MIT Center for Advanced Virtuality (a deepfakes project), reimagines a situation with President Nixon now delivering a speech he never gave: announcing the failure of the Apollo 11 moon landing. The ethical concerns arise from AI's ability to convincingly rewrite history, blurring the line between fact and fiction. The project explores how deepfake technologies can manipulate collective memory and truth by creating an alternative reality. It prompts viewers to question synthetic media's reliability and the dangers of mis/disinformation in shaping public perception. The project includes a physical installation and educational resources, further amplifying its societal impact (Panetta et al. 2019).

Imagine, by French art collective Obvious in collaboration with Sorbonne University, pushes the boundaries of AI by linking functional MRI (fMRI) scans with AI algorithms to reconstruct and generate images directly from human thoughts. This project in neo-surrealism allows AI to visualize the subconscious, enabling subjects to create art based on their imagination and emotions. By transcending text prompts, this work challenges the relationship between human creativity and generative AI, raising ethical concerns around privacy, mind control, and the fusion of human cognition with artificial intelligence (Caselles-Dupre et al. 2024).

The common thread unifying the three artworks is the exploration of the intersection between generative AI and human identity with its perception. *Face Cages* critiques identity confinement, *Moon Disaster* challenges historical accuracy, and *Imagine* explores AI's influence on privacy and human cognition. The ethical concerns in these artworks center on AI's potential to constrain human identity, distort historical truth, and blur the boundaries of privacy and cognition, emphasizing the need for critical reflection on AI's societal impact.

Conclusion

The rise of generative AI presents profound ethical challenges. As AI systems synthesize data-driven outputs without actual cognition, we face questions about assigning moral responsibility to these creations. AI's ability to generate multiple, sometimes conflicting, realities raises critical concerns about truth and relativism. Who decides what's ethically *right* when AI-generated content shapes perceptions? This issue is intensifying with AI's growing influence in shaping the media, communication, and societal narratives, as seen in the ethical dilemmas of business-driven morality and pro-ethical design.

As Stahl et al. (2023) emphasize, there is a significant risk of bias, surveillance, and manipulation embedded in AI systems. Generative AI, while powerful, has the potential to exacerbate societal inequities and human rights violations, accelerating them through generative AI and synthetic media. The need for privacy by design becomes crucial as surveillance capitalism risks undermining fundamental freedoms. This highlights the importance of human-centered AI, which prioritizes ethical responsibility in developing AI technologies, ensuring that they align with societal values rather than corporate interests.

AI's dual-use dilemma—whether it is a tool for liberation or control—further complicates its role in shaping society. Concepts like society-in-the-loop and *trusted AI* aim to bridge this gap by ensuring societal feedback guides AI's development and deployment. The pressing need for integrative ethical frameworks is clear. These frameworks must be flexible enough to manage the fluid ethical context while incorporating utilitarianism, Kantianism, generative AI ethics, inclusivity, and decolonization for responsibility and explainability.

Scholarly Debates

For Cultural and Social Sciences

- How does AI-generated content challenge traditional notions of responsible media practices and human rights?
- How effectively does human-centered generative AI address privacy concerns, prevent manipulation, and mitigate societal inequities?
- How can an integrative ethics framework ensure generative AI development aligns with diverse cultural values, responsibility, and explainability?
- What role does reflective AI art play in uncovering and communicating the ethical dilemmas in media creation and content dissemination?

For Science, Technology, and Engineering

- How can scientists and engineers create synthetic media applications that are explainable in an audit?
- How can innovation support responsibility and mitigate misuse in generative AI?

- What role does generative colonization play in AI development, and how can it be mitigated in a system design?
- How can AI media technologies be developed to respect human rights and prevent discrimination?

For Business, Policy, and Innovation

- How can innovative business projects ensure their AI-driven media practices align with ethical principles and societal expectations?
- How can businesses balance profitability with responsibility and explainability of using generative AI?
- What policy or governance frameworks are essential to ensure AI media's ethical development and regulation?
- What policy guidelines should be enforced in generative content industries if business-driven morality and pro-ethical design drive AI developments together?

Strategic Takeaways

For Academia

- Dual-Use Research: Investigate AI's dual use in media, from social benefits to deception risks.
- Literacy Development: Develop curricula that foster generative AI literacy for responsible and explainable use.
- Multidisciplinary Studies: Integrate insights from technology, social sciences, and humanities to suggest ethical implications for AI media.
- Human Rights and Inclusivity Frameworks: Review ethical-inclusive frameworks for generative AI. Understand their historical and societal roots of bias, inequality, and human rights concerns.

For Industry

- Explainable-Responsible Oversight: Incorporate ethics officers and dedicated committees to monitor generative AI development processes for explainability.
- Diverse Teams: Prioritize diversity in synthetic media development teams to represent a broader range of human experiences and values for responsibility.
- Ethical Investments: Treat generative AI as a strategic investment by balancing business-driven morality and pro-ethical design.
- Inclusive Design: Create synthetic AI systems that focus on social inclusion for equitable access and representation in AI-driven outputs.

For Policymaking

- Regulatory Frameworks: Implement regulations facilitating responsible and explainable generative AI interactions or AI media impacts.

- Social Contract: Apply generative AI as a societal tool bound by ethical agreements between developers, users, and regulators.
- Human-Centered Enforcement: Prioritize human values, needs, and ethical considerations, ensuring that AI media systems improve human capabilities and well-being.
- Society-in-the-Loop: Establish governance frameworks for society in collaboration with generative AI developers, to ensure the systems align with social values, inclusion, and decolonization.

6

POLICY

Defining Policy

In the context of generative AI and synthetic media, policy refers to the frameworks, guidelines, and regulations that govern the development, deployment, and use of AI technologies in content creation, distribution, and moderation. These policies balance innovation and societal values, ensuring AI's impact on media aligns with public interests such as equity, transparency, and sustainability. Key areas include managing intellectual property, authenticity, and algorithmic governance. Co-regulation between public and private entities provides a socio-political power balance between AI, labor, and political economy. Concepts such as data altruism, sovereignty, talent economy, and carbon capitalism underscore the need for socio-environmental consciousness in generative AI practices. The connected artworks critique AI's effects on identity and hyperproduction. Strategic foresight and proactive, rights-based governance can guide the future of AI synthetic content and media.

Introduction

Generative AI and synthetic media are not just evolving and redesigning themselves in a scaling loop; they are shaping fact checks and authenticity, transforming policy and economic structures. As both natural and artificial boundaries dissolve, the focus shifts from control to harnessing their potential to redefine access, power, and opportunity. In the best-case scenario, these dynamics will steer such innovation toward a future that could facilitate human well-being, equity, and societal progress—and several significant forums and platforms have raised the flag for societal AI trust.

However, as we square up to the potential of generative AI, the stakes are high. The technologies involved have the power to radically transform not only

DOI: 10.4324/9781003591023-7

creative industries and media ecosystems but also the political economy and labor markets. Without proactive governance, we risk exacerbating existing inequalities or reinforcing power structures that favor a select few. The capabilities of AI—from producing synthetic media to automating jobs—also demand policies that reflect the fluid and unpredictable nature of technological advancement, where expectations can shift rapidly, and new socio-economic challenges emerge unexpectedly.

The initiation of the G7 Industry and Technology Ministerial Meetings could signal a shift from reactive to proactive policymaking; however, geopolitics may introduce conflicting interests between business policies and governments, and this approach may complicate efforts to establish unified common ground and ignore the potential of wise socio-economic diversity design.

Generative AI is particularly fertile ground for *wild card* events—that is, unexpected disruptions or opportunities that can shift the trajectory of its societal influence. These wild card events have the potential to escalate into *black swan* events—rare, transformative occurrences that fundamentally alter the domain of (generative) AI and its impact on society. As discussed in the chapter "Uncertainties," the interplay between these two types of events emphasizes the need for adaptive and flexible governance frameworks because they can trigger profound and unpredictable shifts in AI capabilities and societal outcomes.

Human-centered approaches are best supported by integrating ethical considerations (see chapter: "Ethics"), sustainable innovation, and equitable access to create a proactive strategy. This multifaceted approach is essential to bring technological investment, implementation, and deployment in real-world contexts into balance. By prioritizing these elements, policies can ensure that AI development is responsible and impactful, addressing societal needs while fostering innovation and inclusivity. This strategy is critical in guaranteeing the alignment of technological progress with long-term societal goals.

This chapter investigates how transformative and generative AI strategies can influence societal values and political systems, and vice versa. It advocates for proactive governance in generative AI policymaking to ensure that technology serves humanity's interests while addressing complex ethical, social, and economic concerns.

Governing GenAI

Since generative AI acts as a co-intelligence (see chapter: "Transformation"), influencing media, political narratives, and economic policies, governance must go beyond traditional regulations. This approach includes both sequential and real-time monitoring, oversight, and evaluation mechanisms to ensure the power of AI is managed responsibly. These frameworks must prioritize explainability, fairness, and alignment with societal values, enabling continuous assessment and accountability.

By embedding ongoing oversight, policymakers can proactively address the ethical and societal challenges that arise as AI systems evolve, ensuring that they operate transparently and equitably in real-world applications. This *proactive strategy* anticipates AI's transformative potential, addressing its societal implications and ethical challenges while balancing innovation with public trust.

In the upcoming quantum computing age, the governance principles outlined by the World Economic Forum over recent years will become pertinent. In particular, if promising future technologies make computational speed and efficiency breakthroughs, generative AI could accelerate rapidly, creating the potential for significant societal and economic imbalances. Such an acceleration could widen the gap between those who control these technologies and those who lack access, posing unpredictable challenges in governance.

The common good, accessibility, inclusivity, and equitability are not merely aspirational benchmarks. These principles echo through the legislative corridors from the European Union to the White House, signifying a concerted effort to control the consequences of the technology, including a safer internet for children and the preservation of free speech without discrimination in democratic contexts. These guiding principles are crucial for ensuring societal and economic balance, forming the foundation of regulatory frameworks amid significantly changing content-driven narratives and evolving information ecosystems. The business insights of consulting firms suggest a future in which human-driven platforms may decline as game-changing GenAI-driven content dramatically rises. This trend signals a monopolized balance if the big tech companies manage the financial challenges posed by the massive computational demands of generative AI, seeing it as a driver of future competitiveness.

The socio-technical perspective advocated by Noorman and Swierstra (2023) provides a holistic lens, marrying the optimization of social and technical systems to improve efficiency and human satisfaction. This approach is central to *self-regulation* discussions in Germany, France, and Italy, emphasizing that risks lie more in AI's application than in the technology itself. This consideration is echoed in the public sector's push for media and generative ethics, where AI's potential for societal benefit must be balanced against the perils of misuse. One alarming example involves AI-generated synthetic media, where entire news reports are fabricated, and false information on global conflicts or war are spread, causing genuine geopolitical tensions. The utilization of detection systems and tools is just one option in mitigating AI-generated mis/disinformation. Establishing guidelines for information platforms is critical to verifying the authenticity of high-risk media and NewsTech.

Co-regulation emerges as a viable governance model. It fosters a collaborative framework somewhere between industry self-regulation and government oversight, ensuring shared responsibility, as Lewis and Moorkens (2020) have described. This model is particularly relevant considering AI's transformative impact on decision-making and widespread application across industries, as summarized by Dwivedi et al. (2021). Regarding AI governance, mediated trust becomes a cornerstone where early ideologies that favored innovation without restraint are now giving way to a critical examination of trustworthiness in technological mediators. In this, Bodo's (2021) theoretical framework on mediated trust urges an integration of internal trust guarantees and external regulatory frameworks. Likewise, Fui-Hoon Nah et al. (2023) argue that the necessity for effective governance and policies becomes undeniably clear in regulating AI's economic and societal impacts.

In data governance, Pavel (2022) calls for a transformative vision, advocating for a digital ecosystem that empowers individuals and serves communities, a vision that aligns with the national self-sufficiency strategies outlined by Gartner Futures Lab (2023). The multipolarity of global politics further complicates this ecosystem, suggesting a move toward regional strategy and models.

As AI assumes a pivotal role in shaping the global media order, Qiu (2023) challenges us to consider neorealist frameworks that focus on states pursuing hegemony in the digital domain. The EU's strategy for data and AI necessitates comprehensive accountability architectures, while the U.S. considers AI safety, privacy, and security.

Robust frameworks become essential for governing generative AI because they address the complex interplay between state-driven technological power and the global influence of synthetic media in the pursuit of geopolitical hegemony. However, as Heaven (2022) highlights, variability in AI decision-making will add further challenges to governance, complicating efforts to ensure responsible oversight of AI's role in shaping global narratives.

Human-Centered Design

A normative, human-centered approach to policymaking allows common ground in tackling the challenges posed by transformative technologies like generative AI. Normative expectations shape public perception and influence the developmental trajectory of such technologies, making strategic foresight indispensable in anticipating future challenges. In line with this, Bingley et al. (2023) and Sigfrids et al. (2023) argue that embedding values like transparency, fairness, and accountability within AI policy frameworks is crucial in ensuring responsible and progressive development.

However, GenAI's impact on creativity, labor, and ethics requires further consideration, particularly in sectors where its applications transform socioeconomic dynamics. A few examples, such as NewsTech, fake news factories, hyperpersonalization, and microtargeting for manipulation, have increasingly challenged reactive public and business policies concerning trusted technologies (see chapters "AI Media" and "Uncertainties"). Transformative and generative technologies blur the lines between legitimate news and fabricated content, raising concerns about media integrity and the ethical use of AI in influencing consumer behavior and political outcomes. The policy implications are profound, as these AI-driven tools shift the dynamics of information dissemination and control, demanding new regulatory approaches.

By fostering proactive governance, policy can transcend reactive approaches, addressing AI's ethical and societal complexities while ensuring its benefits are equitably distributed. Proactive measures could include implementing transparent AI audit systems, regulating the use of generative AI in media through ethical standards, and mandating disclosure of AI-generated content to protect media integrity and public trust while safeguarding against manipulation in political communication and consumer behavior; which thus empower human-centered design in AI ventures beyond the conventional focus on efficiency and economic gain, and beckon us to contemplate a future in which AI serves as a steward of well-being, fostering harmony between humans and their environment. This perspective is not just a lofty ideal but grounded in emerging explorations that challenge us to reimagine the role of AI in societal advancement (Lutz 2023).

This is particularly crucial for generative AI because its capacity to shape content, narratives, and creative processes profoundly influences political discourses, transforms the labor market, and challenges human creative mental health. Therefore, significant challenges are emerging for AI governance, especially considering how GenAI as a co-intelligence holds the potential to dominate key policy and business messages, or political communication narratives. As a consequence, even more complex self-regulatory policy mechanisms are emerging, which require careful oversight to balance innovation with accountability and fairness.

Constitutional AI, for instance, is a framework in which predefined ethical principles, rules, or constitutions govern artificial intelligence systems. This approach further highlights the importance of keeping human values and oversight at the core of AI development. This approach—which has made its way to the steps of Capitol Hill—emphasizes improving human supervision over time, to ensure that AI systems align with our most profound ethical standards and societal norms (Ramani & Wang 2023). This commitment to human-centered AI governance resonates with the broader call for *life-centered design*, proposing a symbiotic relationship between humans, technology, and nature (for the concept of *entangled AI*, see chapter "Uncertainties") in which each amplifies the other's capacity to contribute positively to our world in an intensively interconnected real-time ecosystem.

However, the journey toward a human- or life-centered AI ecosystem is fraught with challenges. The pervasive datafication of society, driven by the imperatives of surveillance capitalism, presents a formidable barrier to this vision. The transformation of family life, democratic processes, and personal privacy under the weight of continuous data extraction and analysis calls for a critical reevaluation of the technologies we adopt and the values they embody (Barassi 2020). Moreover, the discourse around combined content moderation, both algorithmic and human, reveals deep-seated concerns about transparency, fairness, and preserving democratic values in public domains.

The need for a balanced approach that protects individual rights while ensuring collective well-being becomes increasingly complex (Gonçalves et al. 2023) when GenAI or co-intelligence systems are involved. These technologies introduce new layers of challenge in oversight and accountability. The question arises: Who will moderate and define the boundaries of influence in systems where human-AI co-intelligence blends various input datasets, algorithms, and outcomes? Especially if the inputs are a mix of human-AI co-intelligences that contribute to different input datasets, learning algorithms, and outcomes. As GenAI shapes mediatized policy discussions, interests interact within the opaque *black box* of AI, where the decision-making process becomes difficult to trace. This complexity raises questions about trustworthy power, and makes the establishment of precise governance mechanisms to manage the dynamics of the evolution of AI-generated influence without bias or unchecked control an imperative.

The distinction between human and AI interaction or contribution, as Knott et al. (2023) proposed technically, is no longer sufficient to address the growing complexity and far-reaching consequences of GenAI systems. While audits may identify inputs and specific algorithm-driven processes, this reactive method overlooks the need for the proactive, outcome-focused approach that is essential for a genuinely human-centered strategy.

Proactive policies embedding responsibility and explainability into GenAI systems are essential to align AI with societal values and guarantee real-time monitoring. This ensures that generative AI agencies align with societal values and are empowered to shape policy outcomes. Proactive, explainable governance is therefore the key to harnessing GenAI's potential responsibly.

Equity and Human Rights

Generative AI in the content industry and synthetic media introduces innovative opportunities to promote equity and accessibility, but only if governed through inclusive and explainable frameworks. Innovative approaches could involve real-time transparency mechanisms—such as AI-generated content disclosure labels and open-source synthetic media tracking systems—to empower users to assess the authenticity and bias of information. Additionally, collaborative governance models involving civil society, tech companies, and independent auditors can foster shared oversight, ensuring that AI-generated media content is distributed equitably and responsibly, preventing misuse through the amplification of mis/disinformation or the monopolization of cultural data and narratives. These progressive solutions actively address the power asymmetries inherent in AI-driven media and GenAI.

AI-driven media and decision-making systems can either democratize or deepen existing societal disparities. On the one hand, GenAI offers opportunities for broader access to creative tools and content, potentially reducing barriers to participation in the digital economy (Narayanan & Kapoor 2023). On the other hand, concerns arise over data power asymmetries, where large organizations control vast data, thus limiting equitable access to AI's benefits (Pierson et al. 2023). In the media, this imbalance could amplify biases, influencing public narratives and political discourse, as seen in AI's role in detecting political bias in journalism (Wang 2023). A shared governance model that includes diverse stakeholders is, therefore, critical in ensuring fairness, transparency, and accountability in AI's deployment across society (see: *Constitutional AI* above).

A *rights-based governance model* in generative AI supports equity and accessibility by promoting collective data ownership. This would ensure the fair distribution of AI's benefits and shared responsibility for AI-related risks (Lewis & Moorkens 2020). This approach enhances inclusivity by addressing data power imbalances, offering a more democratic framework for generative AI development. Its limits, however, lie in the complexity of implementing collective data rights at scale and ensuring accountability across diverse stakeholders, which may slow down governance processes or create legal ambiguities.

Multidisciplinary research is crucial for establishing responsible and explainable AI governance. Independent academic studies can uncover the intricate relationships between generative AI companies and policymaking, offering a critical lens for evaluating AI's ethical implications. This relational understanding helps validate transparency from a user-centered perspective, ensuring it remains meaningful and trustworthy (Felzmann et al. 2019). For policymakers, engaging with these findings is essential to ensure that transparency in AI systems goes beyond surface-level compliance, fostering accountability and societal trust in AI-driven technologies. The World Economic Forum's recommendations on responsible generative AI resonate with this approach, emphasizing establishing a precise terminology and public awareness, aligning AI with human values, and ensuring accountability and benchmarking. These steps promote global collaboration, social progress, and equitable access to AI, especially in developing countries, fostering diversity and inclusivity in AI development (World Economic Forum 2023).

Despite significant efforts, it is widely recognized that these principles often impose high costs on business development. In some cases, opportunities still need the added influence of activist pressure or regulatory intervention beyond mere narrative.

Human rights initiatives with collaboration between businesses, academia, and policymakers are gaining prominence at leading research institutes and top universities in global AI hubs, advocating for responsible and explainable AI development. This *institutionalized activism* supports the inclusion of minorities, subcultures, and Indigenous peoples in equity and accessibility narratives. While these efforts help mitigate biases and promote a more diverse AI ecosystem, there remains a need for more profound, sustained commitment to ensure that these voices are not tokenized but genuinely integrated into the decision-making process.

The intersection of activism, diversity, and generative technologies raises another crucial concern: *AI colonization.* This refers to the extraction and exploitation of data—often from marginalized communities—to fuel profit-driven AI models, replicating historical patterns of resource extraction and cultural erasure (Hao 2022a, 2022b). GenAI's narrative capabilities further complicate this, as synthetic content generation can replicate and mirror biased and exclusionary narratives. Proactive policy implementation, accordingly, focuses on ensuring AI systems are built with inclusivity in mind, promoting diverse datasets and equitable access to AI-driven opportunities, and also consulting with GenAI activists. These policies also prioritize the protection and explainability of data from vulnerable communities, and foster AI narratives that uplift rather than marginalize underrepresented groups.

AI Power Dynamics

The political economy of generative AI and synthetic content marks a profound transformation in the way economic, cultural, and technological power is concentrated and exerted. By automating labor-intensive processes in content creation, generative AI is fundamentally reshaping industries, particularly at the intersection between creativity, labor, and data. This shift raises critical questions about control, ownership, and the future of work. Central to this transformation is data-intensive capitalism, an economic system driven by data collection, analysis, and exploitation, which is the primary source of profit and competitive advantage (Steinhoff 2022). This system exemplifies how big tech companies, by leveraging massive datasets, maintain their dominance in the global market (Brylska et al. 2024).

Freemium models—where companies offer essential services for free but charge for premium features—play an essential role in this strategy. Such models attract a broad base of individual and institutional users, providing accessible tools while also collecting extensive real-world data for AI training. While this dynamic presents opportunities to focus on human rights and equity issues by incorporating diverse inputs into AI systems, inequalities also may grow dramatically. Marginalized voices may continue to be underrepresented in data collection and design, further entrenching biases already present in these systems (see *colonization* above).

A central connected issue is the concentration of power within *data and AI monopolies*. These refer to a company's control over essential data and AI technologies, limiting competition by monopolizing both resources and innovation pathways, creating barriers to market entry and growth. Thus, companies like Google, OpenAI, Nvidia, Meta, or Microsoft benefit from rentier capitalism. They control the infrastructure and services that power AI technologies, extracting profits from their platforms, the data they collect, and the AI systems they develop. This dominance creates a closed ecosystem in which smaller competitors often become acquisition targets to innovate or compete, and data-driven services can stifle alternative development paths. Legal disputes over licensing and intellectual property occur, evident from recent litigation. These cases add another layer of complexity as companies also grapple with questions surrounding the ownership of AI-generated content (Falyuna et al. 2024).

Equally troubling is the commodification of human interactions through AI platforms. Traditional media models monetized audiences primarily through advertising, but AI platforms have shifted this paradigm. User data is no longer a mere byproduct or source of statistics; it is the core input that drives AI systems. This creates a feedback loop in which user interactions continuously refine AI models, improving their ability to manipulate and monetize user behaviors. This process further deepens the power of platform monopolies, concentrating both economic and cultural influence in the hands of a few dominant firms.

On a geopolitical level, competition between the USA, the EU, and China for dominance in AI governance reflects broader struggles over technological sovereignty. The establishment of AI safety summits, attended by global public and business policy leaders, emphasizes the intense geopolitical race with different strategies. This rivalry transcends technological leadership and extends to the question of who will establish international standards for AI governance, an economic contest that will shape AI's integration into economic and political systems. *Geopolitics of AI*, thus, focuses on AI development and control by various nations and regions that influence global power dynamics, security, and competitive advantage through governance, data sovereignty, and technological leadership.

Beyond the concentration of economic power, AI's environmental implications are significant. Training large AI models is energy-intensive and contributes considerably to the global carbon footprint. Increasingly frequently, policymakers are drawing attention to the ecological costs of generative AI, highlighting the tension between technological progress and sustainability. This issue illustrates a broader contradiction within the political economy of AI, where corporate interests often prioritize short-term profits over long-term environmental stewardship. The ecological impact of AI complicates its governance, requiring careful consideration of how innovation can coexist with climate responsibility. *Carbon capitalism*, as an economic system, integrates the cost of carbon emissions directly into market mechanisms, seeking to reduce greenhouse gases by leveraging financial incentives and penalties.

Synthetic data adds yet another dimension to the *political economy of generative AI*. Unlike traditional datasets, AI systems can generate synthetic data, raising crucial questions about data ownership. Steinhoff (2022) argues that while synthetic data is not entirely divorced from human input, it challenges the assumption that data must originate from human subjects. This shift complicates traditional economic models of data ownership and monetization as machine-generated data becomes a new source of value, further expanding AI-driven capitalism.

These developments collectively challenge traditional concepts of content production. Automating content creation through generative AI goes beyond efficiency gains, marking a shift toward what Ferrari and McKelvey (2023) describe as *hyperproduction*. With this concept, the line between real and simulated content blurs, fundamentally altering both the consumption and production of media. AI models now generate vast amounts of content that reflect human inputs yet are ultimately the products of machine processes. This evolution disrupts the role of human labor and redefines the way in which value is created and extracted from information processes.

Considering all these things, proactive policy implementation of responsible GenAI and explainable synthetic media applies accurate data governance, addresses the growing power imbalances in tech monopolies, and ensures the representation of marginalized communities. These challenge the existing regulatory frameworks that currently conserve monopolies and inequalities. However, if proactive data governance becomes increasingly central to economic and political dynamics then wealth distribution may be transformed in a human-centered way.

Human Labor Market

A critical question emerges at the confluence of generative AI and the evolving labor market: How will the *human workforce*, the human element in labor markets and industries, be redefined in the face of burgeoning artificial intelligence, especially considering co- or autonomous intelligence? Accordingly, the integration of generative AI into workforce management raises four key questions. First, it prompts an examination of the time investment required for tasks performed without AI assistance. Second, it calls into question compensation for those who currently perform such tasks. Third, the accuracy with which AI can execute these tasks is evaluated. Finally, attention is given to how easily human workers can verify the accuracy of AI-generated outputs (MIT IDE 2023a, 2023b). These criteria serve as a well-designed decision-making compass for those organizations contemplating the transition from human to artificial intelligence labor, acknowledging that the true efficacy of generative tools is already available for complex tasks.

The transformative tech economy, fueled by expanding digital infrastructures and the rise of remote work, offers an increase in opportunities and types of access for some populations, advancing both workforce participation and productivity. However, this shift brings both advantages and challenges. While digital access can empower diverse talent pools and reduce geographic barriers, it may also disrupt existing labor structures, creating unequal access to resources and skills. The impact of new market entrants, particularly in generative AI and synthetic media, will depend on how organizations and policymakers address the rise of AI-generated content, by ensuring regulatory frameworks are in place to manage disruptions in creative industries, intellectual property rights, and inequalities in access to these technologies.

The future of work, moreover, is being reimagined as human-centered, with a blend of flexible work experience and empathetic management. This evolution necessitates adapting those work models that integrate life-centered design with central technologies like generative AI. As Riedl (2019) argues, human-centered AI is twofold: AI systems must understand humans from a socio-cultural perspective while helping humans understand how machines work. This synergy can ensure that AI functions effectively within the broader system of human interactions, fostering a more collaborative work environment. As organizations strive to meet specific employee priorities—fairness, equity, autonomy, flexibility, and social justice—the competition for talents, especially those skilled in AI arenas, remains fierce. As the Gartner Futures Lab concluded, the rise in labor organizing within the tech industry underscores the persistent talent crunch despite high-profile layoffs (2023). Highly skilled GenAI professionals and well-designed services are notably absent from policy implementation. However, policy initiatives can address this gap by upskilling talent through governance models and *learning by doing* approaches, fostering practical expertise, and engaging the new generations in human-centered AI. The model of *talent economy* supports this direction, where growth is driven by skilled individuals, with policies focusing on developing, attracting, and retaining talent to foster innovation and address evolving industry demands.

The rapid integration of generative AI into the labor market presents both opportunities and significant challenges. While GenAI transforms industries like media and content creation, it also introduces critical issues such as biases, the spread of AI-generated mis/disinformation, and the reshaping of public opinion (Ananny 2024a, 2024b). These developments complicate transparency and accountability, creating a significant need for ethical frameworks that guide responsible AI usage. Policymakers face the difficult dual task of mitigating these risks while ensuring GenAI remains a catalyst for innovation, rather than a source of societal harm.

Strategic foresight is key in addressing these challenges. It enables the development of adaptive policies and ensures resilient decision-making in response to evolving labor market dynamics. The Artificial Intelligence Center of Excellence (2023) emphasizes the importance of structuring organizations to accommodate AI, by deploying it ethically and developing a skilled AI workforce. This approach aligns with the labor market's need for inclusive and responsible AI applications.

A growing industry of responsible AI practitioners and experts reflects this shift, especially in global AI hubs. Feher et al. (2025) highlight this trend in their research on hands-on strategic foresight for responsible AI futures, finding common ground and cross-industrial consensus among stakeholders via specific strategic foresight methods.

As GenAI redefines job roles and fuels demand for specialized talent, governance structures and upskilling initiatives gain strategic importance. The evolving talent economy challenges traditional labor models, pushing policymakers to reconsider workforce dynamics. This includes assessing the costs and benefits of workforce development, particularly as AI and human labor work together to transform productivity.

Creativity and Social Innovation

When AI-driven creators first received awards and presented their technological collaborations, critics primarily shaped an offensive discourse that echoed the initial reactions to avant-garde and readymade art. However, as AI has increasingly integrated into the creative industries, it has not only transformed traditional concepts of authorship and artistic expression but also driven social innovation.

Generative AI transforms creativity by extending beyond human limitations and blending human imagination with algorithmic power in particular. Batori (2024) argues that AI-driven content manipulation does not inherently deceive or mislead. Instead, artistic freedom is discussed, where such manipulation enhances creative expression rather than distorting reality. This shift challenges current policy frameworks, emphasizing the need for new approaches to intellectual property, creative labor, and fair access to AI tools in the cultural sector, to promote both artistic and social innovation.

The impact of generative AI on creativity is far-reaching, affecting everything from visual arts to literature, music to architecture. Its algorithms, such as those driving text-to-image generators or composing music, have transferred the creative process. They offer many options, iterations, and variations at a speed and volume unattainable by humans alone. This surge in creative output, while exhilarating, also raises profound questions about the nature of creativity itself. In the news industry, for instance, creative brainstorming and illustration production represent the exploratory phase in GenAI, while efficiency or cost-effective solutions are in second place (Thomson et al. 2024). This ratio can change with the human learning process in GenAI and the human-feed GenAI, leading to a more efficient collaboration where human insights refine AI outputs and AI accelerates human creativity, especially as the communication and media industry's reliance on technology further accelerates news and information distribution cycles (Zelizer 2021).

Social innovations facilitated by generative AI are equally groundbreaking. AI-driven platforms are able to transform previous communication forms, democratizing content creation and dissemination. These platforms enable users from diverse backgrounds to share their narratives, fostering a more inclusive media ecosystem. Moreover, generative AI has the potential to address complex societal challenges. For instance, AI can generate simulations to predict social patterns or crises, aiding policymakers in crafting more effective interventions. It can also curate personalized educational content, by adapting to individual students' learning styles and needs, thus transforming educational trends.

However, the rise of generative AI complicates the distinction between human and AI authorship, raising challenges regarding *intellectual property rights*. Stock libraries banning GenAI requests and educational institutions blocking AI in creative courses illustrate how copyright law is struggling to keep pace with technological advancements (Mantegna 2024). Bans, on the other hand, cause resistance and create countertrends. Proactive regulations and policy incentives can be proposed to prevent this in adapting to the emerging GenAI. These can shape future-oriented originality and credibility for foresight-driven intellectual property norms, keeping human creators in the AI scaling loop.

However, the democratization of creativity through AI is a double-edged sword. While it levels the playing field, allowing individuals with limited resources to participate in creative endeavors, it also challenges traditional creative industries and professionals. The economic implications for artists, musicians, writers, and others in creative professions are profound, necessitating a re-examination of the value and compensation of innovative work in an AI-augmented market. Creativity and social innovations driven by generative AI not only reshape job roles and workforce dynamics but also demand progressive governance structures and upskilling initiatives that ensure inclusive growth, balancing AI's transformative impact on productivity with equitable opportunities for human talent development.

GenAI for Good

Promoting social, environmental, and economic benefits—including healthcare, education, sustainability, and humanitarian efforts—is an example of *AI for Good*. This approach is critical as it guides responsible AI deployment, ensuring technologies align with ethical standards and contribute positively to global welfare and equity.

Generative AI can also play a transformative role in policy for *social good*. GenAI can explore its impact on ecological sustainability and socio-cultural heritage preservation via synthetic content. In environmental policy, AI-generated media enables the simulation of ecological effects and supports sustainable infrastructure design through data-driven visualizations. Synthetic contents can reconstruct historical and immersive experiences for socio-cultural heritage, preserving several endangered languages and traditions.

GenAI can ensure inclusivity in these domains by democratizing access to natural and socio-cultural resources. These innovations can enhance policy frameworks by promoting sustainability, cultural resilience, and social equity, addressing the needs of vulnerable communities while supporting long-term societal well-being. As the OECD AI Principles outline (2024), policy instruments and strategies offer a structured approach to the implementation of AI. These policies are crafted to harness AI's potential while addressing ethical, economic, ecological, and social challenges. Focusing on sustainable AI strategies ensures that AI developments align with broader societal goals and handle the risk of resource depletion (Brevini 2021). The most representative global initiative in this field is the United Nations' "AI for Good," which leverages AI to advance the Sustainable Development Goals, addressing issues like poverty, education, and sustainability. By fostering responsible use of AI, promoting equity, and supporting global cooperation, GenAI benefits humanity, particularly underserved communities, for social good. Collectively, these frameworks define the concept of social generative AI responsibility, centering on the ethical use of generative AI to advance sustainability, fairness, transparency, and inclusivity. This includes addressing challenges such as the detection of synthetic outputs, content manipulation, disinformation, algorithmic bias, and ensuring that GenAI-driven innovations align with societal values while minimizing potential harms and fostering positive impacts across industries.

Data altruism, in line with this, promotes the voluntary sharing of data for the public good. This approach fosters data reuse, providing a framework with clear rules for data intermediaries while redefining traditional notions of data ownership. Data altruism enables AI systems to leverage diverse datasets for societal benefits. By positioning data as a shared resource, this paradigm shift supports policies that promote equitable access and explainable AI use for sustainable development and social equity (Pavel 2022).

On the flip side, *data sovereignty* introduces a competing concept, as nations prioritize control over their data assets to safeguard autonomy and enforce local regulations. While this approach enhances national governance, it can restrict global data sharing, complicating efforts to create cohesive AI policies.

Data sovereignty and data altruism are competing concepts, which is especially evident in the rise of synthetic media in NewsTech. Nations seeking to assert control over AI-generated content may resist global regulatory efforts, leading to significant tensions. This resistance heightens debates around mis/disinformation, media integrity, and the cross-border flow of AI-driven news. High-profile cases, such as deepfakes in political campaigns or AI-generated news content, underscore the challenges of balancing national interests with the need for cohesive international AI governance, further complicating efforts to establish universal standards for GenAI.

AI's role in championing social good is gaining traction. Projects are evaluated based on their ability to address and solve societal issues without exacerbating existing problems. This movement toward "AI for Social God" seeks to leverage technology for sustainability and a reduction in inequalities (Tsamados 2021). However, the transition to AI-driven solutions is challenging. The rapid integration of AI in education and other sectors has led to an illusion of transformation, which means the need for a more nuanced understanding of AI's impact on learning processes and societal outcomes is often bypassed. Education policies are increasingly turning to AI as a panacea for systemic challenges. This trend is met with caution by education sociologists, who warn against a deterministic view of AI's capabilities. The integration of AI in education requires a balanced approach, one that is informed by empirical research and societal needs (Davies et al. 2021). As GenAI reshapes how knowledge is conveyed and accessed, *education policy* also plays a transformative role. A balanced, research-informed approach is necessary to ensure students can effectively learn and engage with knowledge in the age of Generative AI, without over-relying on its capabilities.

The legal principle of *fair use* allows limited use of copyrighted material without permission, for purposes such as education or research, considering factors like purpose, amount, and market impact. AI ethics, accordingly, advocate for responsible use and explainable outputs, by engaging diverse stakeholders and formulating policies that address potential social impacts (Stahl & Eke 2024). This ensures that GenAI can be leveraged ethically while respecting intellectual property and fostering social good.

Nature Portfolio's decision to limit AI-generated imagery reflects growing concerns about AI's impact on research integrity and intellectual property. The potential for AI to create misleading or inauthentic content raises crucial questions about maintaining transparency and trust in scientific publications. This highlights the challenges of formulating regulatory frameworks to ensure AI is used responsibly and preserves research integrity (Nature 2023). Moreover, suppose generative AI monopolies exploit copyrighted scientific publications without consent. In that case, they risk disrupting the equitable distribution of knowledge, potentially prioritizing profit over the common good (Sartori & Theodorou 2022) and undermining scientific integrity.

As we envision the future of AI, it becomes evident that AI's role in society is an evolving narrative in expectations regarding social good. With one example, the Insight Forums in the U.S. Senate and the varied responses to AI's integration underscore the need for policies that foster ethical and sustainable AI development, to ensure that AI's potential is harnessed for all (Miller 2023). With another example, the calls for international AI standards thus reflect the need for global cooperation to address the challenges posed by AI (Zakrzewski & Faiola 2023, Zakrzewski et al. 2023). These approaches address the democratic deficit and sustainability issues in AI development and deployment, ensuring that the technology aligns with shared values and effectively addresses societal concerns (Schaffer 2022, 2024).

Contrasting AI Policies and Strategies

The global arena of AI policies reflects a tapestry of diverse strategies as nations seek to harness the power of artificial intelligence. One key consideration is the role of big tech in shaping power dynamics, raising essential ethical and political questions (Coeckelbergh & Gunkel 2023). The other is explainable AI (XAI) playing a crucial role in shaping policy, particularly in regulatory frameworks, where transparency and accountability are paramount.

For instance, the EU's General Data Protection Regulation (GDPR) mandates the right to an explanation in automated decision-making. At the same time, the EU AI Act (European Union 2024/1689) emphasizes risk assessment and accountability in AI systems, and the Digital Services Act (European Union 2022/2065) focuses on transparency and user protection in digital services, ensuring ethical use of AI. The U.S. Executive Order on AI (White House 2023/14110) prioritizes safe, accountable AI across federal agencies, while

California's AI Transparency Act (California Legislature 2023/AB 331) addresses GenAI specifically, requiring detection tools, content disclosures, and license compliance. In contrast, China's generative AI regulations enforce alignment with socialist values, filter sensitive topics in real time, and ensure "politically correct" responses for national security (Hu 2024).

These frameworks represent a comprehensive strategy for AI oversight, emphasizing transparency, accountability, and explainability while accommodating diverse cultural and policy preferences. A similar regulatory trajectory is anticipated in the future, ensuring ethical progress while driving varied cultural and policy imperatives. These frameworks face challenges through ISO standards for generative AI if the goal is to harmonize regional regulations with international standards, thus ensuring privacy, security, risk management, and ethical compliance, balancing local-regional priorities and addressing global transparency, accountability, and explainability.

XAI ensures compliance with such regulations and equips policymakers with tools to create guidelines promoting responsible AI usage, enabling the public to better understand AI technologies' impact. However, while aiming to prevent misuse, this regulatory approach may slow innovation by imposing stringent requirements. For example, in Generative AI, a regulation framework could limit advancements by enforcing strict disinformation controls, potentially delaying creative applications in media and information ecosystems. Policy incentives are critical as a counterbalance—such as supporting synthetic media for educational and creative purposes while mitigating its misuse for disinformation.

Recognizing AI's transformative potential, national AI strategies are now widespread and strongly emphasized in developed nations. Many countries are shaping their AI futures, prioritizing privacy and data protection. The MENA region, notably through Palestine's national strategy, also shows increasing commitment to AI-driven development (Demaidi 2023). This global momentum highlights the growing significance of AI governance, transparency, and equitable access, clearly focusing on leveraging AI's economic potential, including its impact on GDP and broader social development. The economic impact of AI, forecast to boost global income by up to 40%, underscores the urgency of robust and scalable data infrastructures. Advanced systems like data lakehouses, which combine data lakes and warehouses, are reshaping the economic environment and propelling nations into a transformative era of digital democratization (MIT Technology Review Insights 2023).

In AI-driven media, social platforms face evolving business models being under growing ethical scrutiny. Legislative hearings, privacy concerns, and the search for sustainable revenue models all challenge the integrity of user data. This transformation, aligned with global AI governance trends, highlights the tension between innovation and ethical responsibility (Brown 2021). AI's role in content moderation and hyperpersonalization presents new opportunities and challenges, requiring platforms to balance profitability with user rights and trust. As significant contributors to GDP, these media businesses are pivotal in shaping policy recognition, further intertwining economic interests with regulatory frameworks.

The *Beneficial AI Roadmap* distills AI governance into four cornerstone goals: investing in existing AI applications, bolstering trustworthy AI research, democratizing AI oversight, and monitoring AI development. These goals advocate for a future where AI advances safety, inclusivity, and democratic values without succumbing to the trade-offs that can stifle innovation (Bengio & Privitera 2023). Contrasting the U.S. and EU, the policy approaches are partly different. In the United States, a sweeping executive order signals a commitment to embedding AI across federal agencies, streamlining high-skills immigration, and establishing oversight of the private sector's AI developments. This directive aligns the nation's economic, health, and educational sectors with the AI transformation, ensuring a monitored and transparent integration of AI technologies (Hatterjee & Kern 2023). In Europe, the Commission's persuasive discourse has effectively aligned Nordic states with a coordinated AI policy framework, showcasing the role of transnational actors in national policymaking (Malmborg & Trondal 2023). While various nation's efforts at harmonization converge on the narrative of AI's disruptive inevitability, they diverge in their unique cultural, political, and economic responses to AI's integration and regulatory challenges (Bareis & Katzenbach 2022).

These policy strategies address the governance of Generative AI within a complex geopolitical environment. As GenAI reshapes industries and social dynamics, international collaboration can mitigate risks and manage uncertainties. AI governance interoperability, emphasizing coherent coordination of systems, policies, and ethical standards, could enhance transparency, manage mis/disinformation, and secure intellectual property across borders. This pragmatic approach aims to guide AI development toward a global common good while respecting national priorities and data sovereignty. Without proactive policies, GenAI risks widening digital divides and amplifying global inequalities, underscoring the need for strategic foresight and accountability in regulation.

Impact Project

The European Digital Media Observatory (EDMO), launched in June 2020, was established to create a cross-border, interdisciplinary network designed to improve Europe's collective response to online disinformation. Over the past four years, EDMO has made significant strides through its work in fact-checking, media literacy, research, and policy analysis. As generative AI technologies rapidly advance, including their potential to generate vast amounts of inauthentic content, EDMO has recognized the need for a policy-driven approach to mitigate the risks AI poses to the integrity of the media and democratic processes.

Generative AI, while offering technological advancements, presents a critical challenge: it enables the production of disinformation at scale, complicating the public's ability to distinguish between credible and fabricated content. This presents a unique challenge for policymakers, particularly as AI technologies advance faster than regulations can keep pace. EDMO's response comes at a crucial time, as the AI Act, although adopted, was not yet effective in regulating AI's influence during the 2024 EU elections. This regulatory gap underscores the urgency of EDMO's initiatives to fill the void through proactive measures to safeguard the media ecosystem and protect democratic processes.

Since March 2023, the EDMO has taken concrete steps to address AI-generated disinformation through its Monthly Fact-checking Briefs, offering regular insights into AI-driven disinformation trends. In addition, creating a multidisciplinary Task Force in January 2023 that comprised of researchers, fact-checkers, and media literacy practitioners exemplifies a coordinated approach to integrating research and policy action. The Task Force's findings, particularly regarding AI-generated audio content, reveal that current policy frameworks must evolve to address the nuances of AI disinformation that can easily evade detection.

In September 2023, the EDMO strengthened its policy-oriented approach by establishing an Expert Group on Disinformation and Generative AI. This group assesses the risks posed by AI-driven media and explores how policy can support the development of AI-powered technologies to detect and mitigate disinformation. The insights and recommendations from this group will be pivotal in shaping future EU regulations and AI governance frameworks as policymakers increasingly recognize the need for adaptable and forward-thinking policies that anticipate AI's evolving capabilities.

From a long-term policy perspective, EDMO's initiatives set a critical precedent for how Europe can respond to AI-driven disinformation. While current AI-generated disinformation accounts for only 4%–5% of detected false content, the rapid advancement of AI technologies—coupled with user reliance on AI systems like Large Language Models (LLMs)—poses significant risks to public trust and democratic stability. EDMO's work, particularly in the context of the Digital Services Act (2024), illustrates the importance of integrating policy, technology, and public awareness to address these challenges comprehensively. By proactively engaging in policy-driven initiatives, EDMO is mitigating immediate risks and shaping the long-term regulatory framework needed to ensure that AI promotes, rather than undermines, media integrity and democratic processes.

Reflective Art

Artworks inspired by policy for generative AI and AI media serve as crucial touchstones for policymakers, providing them with alternative lenses to comprehend the multifaceted impacts of these technologies. They bridge abstract policy discussions and tangible human experiences, translating complex concepts like data altruism, platform governance, and AI's societal implications into visceral, engaging formats. Such creative endeavors demystify and disseminate policy issues to a broader audience and provoke critical thinking and reflection. They offer a mirror to society, showcasing potential futures shaped by today's AI strategy and governance decisions. In doing so, these artworks catalyze a more informed dialogue around the development and regulation of AI, urging a balance between innovation and regulative frameworks.

Spatial journalism from The New York Times R&D (2021) explores how spatial computing technologies can enhance storytelling by embedding news within immersive, location-specific synthetic experiences. This project leverages AI and spatial data to create context-aware, generative narratives, making it policy-relevant as it highlights privacy, data ethics, and accuracy in journalism.

The Saint George on a Bike project, led by the Barcelona Supercomputing Center (2022) and Europeana Foundation, uses AI and the MareNostrum 4 supercomputer to analyze Europe's vast cultural heritage. By generating automatic descriptions of historic images, the project aims to promote access to and public understanding and awareness of cultural heritage. Through crowd-sourced annotations and AI's contextual adaptations, *Saint George on a Bike* enriches metadata for educational, social, and historical applications, fostering a more inclusive European identity.

The Ford Foundation Gallery's *What Models Make Worlds* (2023) involves *Critical Imaginaries of AI* that showcase artists exploring AI's societal implications through themes of justice, inclusivity, and transparency. Curated by Mashinka Firunts Hakopian and Meldia Yesayan, the exhibition critiques the biases embedded in AI systems—highlighting works that confront surveillance, data erasure, and algorithmic discrimination.

These artworks act as vital conduits for policy feedback, offering nuanced insights into the complex interplay between technology, society, and governance. Artists, often at the forefront of technological experimentation, have engaged with AI since long before its widespread applications and consequential impacts began to materialize. As Cook and Palankasova (2023) emphasize, through their critical reflection on AI, these works thus transcend conventional dialogues, providing a deeper, more embodied understanding of AI's implications for equity, sustainability, and human life. Consequently, these artistic explorations reinforce the importance of developing inclusive and contextual policy.

Conclusion

Generative AI and synthetic media are not simply evolving technologies but actively reshaping policy frameworks and economic structures. Blurring natural and artificial boundaries can shift the focus from merely controlling these innovations to harnessing their potential to redefine power, access, and opportunity in society. In an optimal scenario, these dynamics could advance human well-being, equity, and societal progress, supported by platforms that emphasize AI's societal trustworthiness.

GenAI's influence extends beyond the creative industries, impacting the political economy and labor markets. Its potential to reinforce existing inequalities or exacerbate power imbalances calls for nuanced regulatory approaches. Governance must account for the fluid, unpredictable nature of AI-driven technologies, where technological progress can transform societal structures. Initiatives such as high-level technology policy meetings mark a shift toward proactive policymaking, yet conflicting interests complicate efforts toward the establishment of international standards. The geopolitical arena influences such negotiations, revealing the need for policies that consider socio-economic diversity while mitigating risks.

By focusing on human-centered design and integrating sustainable innovation, equitable access, and ethical considerations, policies can align AI development with broader societal goals, fostering both innovation and inclusivity while balancing technological progress with social responsibility. GenAI policy, therefore, can be structured to govern generative AI technologies comprehensively, addressing key areas such as intellectual property, mis/disinformation, social impact, and data altruism vs. sovereignty while also considering AI's ecological footprint and its role in promoting sustainable practices.

Scholarly Debates

For Cultural and Social Sciences

- How can policies ensure that AI-driven media supports creativity and social innovation without compromising ethical standards?
- In what ways do AI governance frameworks impact equity and accessibility in synthetic media?
- What role does policy play in balancing intellectual property rights with fair use in AI-generated content?
- How can policy frameworks ensure that GenAI serves both social and environmental good?

For Science, Technology, and Engineering

- How can AI systems be designed to comply with policy guidelines on intellectual property and content ownership?
- How do policies encourage human-centric AI systems for media applications?
- How can GenAI policies promote sustainable technological innovation while addressing environmental concerns?
- What challenges arise when aligning GenAI's technical advancements with regulatory frameworks across different countries or regions?

For Business, Policy, and Innovation

- How can businesses manage the complexities of AI policies to ensure compliance while promoting innovation in the content industry and media?

- How can businesses leverage GenAI policies to foster creativity and social innovation while ensuring environmental sustainability?
- What strategies can policymakers employ to balance AI-driven innovation with protecting intellectual property and fair use?
- How can AI governance policies shape global business strategies and ensure equitable access and collaboration?

Strategic Takeaways

For Academia

- Governing GenAI and Sustainability: Investigate human-centered design for sustainable generative AI systems that effectively integrate with environmental frameworks.
- Human Rights in AI Power Dynamics: Examine how GenAI affects human rights, focusing on equity, accessibility, and power imbalances in AI systems.
- Creativity and Social Innovation: Explore how synthetic media improves creativity and social innovation while balancing technological progress with policy needs.
- Disinformation Governance: Analyze AI policies combating disinformation in the media, focusing on truth, accountability, explainability, and equitable content moderation.

For Industry

- Proactive GenAI Strategy: Prioritize well-being and sustainability in generative AI beyond profit-driven and operational goals.
- Constitutional AI Oversight: Ensure GenAI aligns with social good through co-regulation and self-regulation frameworks.
- Talent Economy and GenAI: Utilize generative AI in talent management to foster innovation and gain a competitive edge.
- Freemium Content Moderation: Implement GenAI content moderation, which offers essential free services for fairness along with premium features to ensure explainability.

For Policymaking

- Proactive GenAI Strategy: Develop generative AI policies using co-regulation to ensure societal benefits and responsible technological progress.
- Human-Centered GenAI Governance: Implement rights-based governance to align generative AI with human rights, autonomy, and social integrity.
- Sovereignty vs. Altruism in NewsTech: Balance data sharing with control to prevent monopolies and ensure media integrity in AI-driven news.
- GenAI and Carbon Capitalism: Regulate generative AI's role in carbon capitalism to ensure sustainability while encouraging innovations.

CONCLUSION

This work transcends the hype on generative and synthetic AI tools and their surrounding general concerns or temporary regulative approaches. Instead, it captures these developments as critical milestones in technological evolution, revealing their far-reaching societal impact. By focusing on rigorous analysis, strategic foresight, and intentional policymaking, this book invites readers to consider how generative AI is at a societal scale affecting public perception, societal-business-environmental uncertainties, ethics, and policy across sectors. This perspective supports the "Societal Scale of GenAI," where the pervasive influence of AI-generated content transforms interactions, norms, and policies.

Generative AI outputs have become indispensable across various sectors, raising critical questions about intellectual property, authenticity, and creativity. These technologies transform how content is produced, consumed, and trusted; and they challenge established notions of originality, accuracy, and reliability. The distinction between authentic and synthetic content is increasingly blurred as generative AI draws on vast, often unverified global datasets that constantly change in real time. Without early and consistent identification of AI-generated content, the line between human- and machine-produced work may vanish, further complicating or overriding knowledge integrity and societal structures.

Accordingly, static detection systems, while helpful, are already inadequate for real-time recognition of AI-driven content. More advanced, adaptive methods—such as real-time verification, context-aware detection, explainability, blockchain tracing, and quantum encryption—will be crucial in upholding ethical standards. These compliances can ensure transparency, particularly as policy frameworks evolve to proactively accommodate generative AI's rapidly expanding capabilities and continuous fine-tuning.

The uncertain nature of generative AI's evolution adds complexity to its governance. Its capacity to learn, adapt, and evolve outpaces traditional

DOI: 10.4324/9781003591023-8

regulatory frameworks, leaving them outdated. Static policies cannot effectively manage a technology that develops in real time. This book advocates for a dynamic governance approach, emphasizing the necessity of strategic foresight—that is, methods that predict future challenges and risks. Without such adaptive policies, society remains reactive, addressing only the symptoms of AI-driven disruption rather than its root causes. These root causes include algorithmic bias, data privacy concerns, and lack of transparency.

Yet, amidst these challenges, generative AI also presents profound opportunities. AI has the potential to democratize creativity, scale content production, and offer new ways of engaging with media. However, to harness this potential, society must ensure that the benefits of generative AI are equitably distributed. This requires not only technological innovation but a commitment to social equity. The book has explored the idea that AI's integration into media and content industries is not inherently neutral; it can exacerbate inequalities or offer new pathways to inclusion. The future of AI in the media will be defined by how society chooses to govern it—whether AI serves the public good or risks societal stability.

The key policy implication is co-regulation, where public institutions collaborate with private entities and academic research. This collaboration is essential for addressing the socio-political power dynamics that generative AI introduces. Concepts like data sovereignty, carbon footprints, and data altruism offer glimpses into how policymakers might address the ethical and environmental concerns that come alongside the rise of generative AI. However, policymakers can also take a more proactive approach by embedding strategic foresight into their governance models to ensure that AI serves societal interests over the long term.

The future of generative AI and AI-driven media depends on society's ability to manage its complexities, particularly through AI governance interoperability, which ensures seamless coordination between systems and policies. This approach aligns the generative AI value chain across its lifecycle—from data collection to content creation—while emphasizing continual feedback-driven improvements. Ultimately, this supports social generative AI responsibility, promoting fair, trusted, and explainable generative AI futures.

This monograph addresses the transformative challenges posed by generative AI, which increasingly blur the lines between fact and fabrication in human-machine interactions. This transformation should raise innovative and future-oriented approaches to authenticity, credibility, and the very nature of information. The rapid synthesis of diverse data magnifies this transformation and represents immediate societal challenges, no longer abstract concerns. Thus, forward-thinking, intergenerational experience sharing is critical in ensuring

With the emergence of Agentic AI, that generative AI fosters societal stability. as the augmented evolution of generative AI, this transformative approach is becoming increasingly critical. Moving beyond passive transformation, it enables proactive decision-making and adaptive goal-setting, signaling a significant technological advancement (Randieri 2025). While Agentic AI represents a promising innovation, its autonomy introduces even more complex uncertainties, underscoring the need for progressive and adaptive governance to address societal challenges effectively.

Considering this AI evolution, this book contributes to the emerging and critical intersection between academia, business, and policy by offering foundational insights into these frameworks. It aims to prepare decision-makers to engage with the profound technological transformations reshaping society. Concluding the book, three central questions remain:

- *UNCERTAINTIES*

How do we govern generative AI's unpredictable evolution without stifling its creative potential?

- ETHICS

How can co-intelligence in generative AI drive ethical growth while preventing unintended societal consequences?

- POLICY

How can policies evolve to keep pace with generative AI's real-time learning, avoiding constant reactionary updates?

- THE ULTIMATE QUESTION

What are the limits of generative AI's transformative potential for scaling in society?

- FINAL CONCLUSION

The future of generative AI is not just about implementing a promising technology—it's about transforming society to align with the potential of generative AI.

GLOSSARY

5G	A wireless technology offering significantly fast speeds and reduced latency, enabling real-time data processing and enhanced connectivity across devices and applications.
6G	An advanced wireless technology that delivers ultra-fast speeds, near-zero latency, and intelligent network capabilities supporting real-time immersive experiences and the seamless integration of AI-driven applications.
Accountability	The concept ensures that decision-makers and AI developers are held responsible for the outcomes of AI systems, particularly in ethical, legal, and social contexts, by defining clear roles and consequences.
Agent/agency	The capacity of humans or AI to perceive, decide, and act autonomously to shape outcomes.
Agent behavior	Autonomous decision-making and actions of AI systems shaped by programming relational ethics, with impact on social norms, media, and human interactions.
Agentic AI	This emerging concept refers to autonomous systems that transition from passive responses to proactive goal-setting, decision-making, and adaptation to changing environments, surpassing the capabilities of reactive generative AI.
AI agent	Autonomous system that interacts with its environment, perceives inputs, and makes decisions based on its programming and objectives to achieve predefined goals.

AI development imperative	Advance AI technologies and their applications to address complex problems and enhance various aspects of life and industry.
AI drift	When AI systems gradually deviate from their original goals or intended functions due to data, environment, or interaction changes.
AI-driven AI/VR/XR	AI integration with AR, VR, and XR enriches immersive experiences through intelligent interaction, dynamic content creation, and personalized engagement.
AI fatalism	The belief that AI-driven outcomes are inevitable, reducing human agency and fostering deterministic views on technology's societal impact.
AI for Good	Promoting social, environmental, and economic benefits—including healthcare, education, sustainability, and humanitarian efforts—for positive societal impact.
AI-generated magazine cover for *Cosmopolitan*	A human-GenAI collaboration art project challenging traditional creativity and authorship and raising uncertainties about AI's role in design professions and the authenticity of creative works.
AI generators	Tools or systems that use AI algorithms to produce synthetic outputs automatically.
AI governance	The dynamic systems of policies, regulations, and ethical frameworks that shape AI's development, deployment, and impact to ensure accountability and societal alignment.
AI literacy	Understanding the fundamentals of artificial intelligence, including how it works, its applications, and its societal impact, which enables individuals to engage with AI responsibly and critically.
AI media	News, entertainment, social, virtual, synthetic media, and "media as a service" are driven by data analytics and generative technologies, impacting production, distribution, and moderation.
AI media ethics	The responsible application of generative technologies in media, focusing on their societal impact and transformation of creative processes through advanced production, distribution, and innovative media practices.
AI media future	Advanced media production and services powered by recognition-behavioral-transaction technologies and sophisticated analytics.

AI mediatization	Integrating advanced AI technologies into media processes, transforming traditional ecosystems by improving automation, disrupting workflows, and adding new layers of complexity to media creation, distribution, and consumption.
AI strategy	A plan or approach that outlines how an organization or individual intends to leverage AI technologies to achieve specific goals or solve problems.
AI Surrealism Exhibition	This was the world's largest AI art exhibition. It emphasized AI's provocative role in art while promoting sustainability through AI-generated nature imagery to build connections and encourage environmental awareness.
AI-to-AI loop	The integration of AI technologies, where AI systems train each other and enhance overall functionality, leading to improved performance and innovative applications.
AI winters and springs	Phases characterized by reduced funding and interest in AI research and development (winters) followed by revitalized enthusiasm and significant advancements (springs).
AI's dirty mind	Refers to the tendency of AI systems to generate inappropriate, offensive, or unexpected outputs due to biases in training data or flawed algorithms.
Algorithmic awareness	Understanding how algorithms function and shape behaviors, especially in generative and mediatized environments, while recognizing the effects of automated systems on personal choices, privacy, and broader societal consequences.
Anthropomorphic framing of generative AI	Attributing human-like traits and emotions to GenAI systems, influencing public perception, trust, and ethical considerations regarding the capabilities and responsibilities of the technology.
Anticipatory infrastructures	Systems utilizing AI, data analytics, and sensors to predict and address future needs, enabling proactive decision-making and adaptability in response to emerging societal, environmental, and technological challenges.
Artificial intelligence (AI)	The simulation and augmentation of human intelligence processes by machines, especially computer systems, which includes learning, reasoning, and behavior.

Artificial-Synthetic-Hyper-Immersive Media Extensions (ASHIMEs)	An advanced artificial environment designed to deliver intensely immersive experiences by integrating virtual and augmented reality with other transformative, algorithm-driven technologies to enhance human perceptions, interactions, and media experience.
Attention economy	The concept that human attention is a scarce and valuable resource that businesses and platforms compete for—often using AI-driven algorithms to capture and retain user focus, influencing behavior and decision-making.
Authoritative style of GenAI outputs	The confident and often definitive tone of AI-generated responses contrasts with experts, who recognize the limits of their knowledge and embrace uncertainty, highlighting the gap between AI's perceived certainty and human expertise.
Autonomy	The term refers to the ability of a system or entity to operate independently, make decisions, and perform tasks without external control. In transformative applications, autonomy is often driven by advanced algorithms or AI.
Bias	The term refers to systematic errors in data or algorithms that lead to unfair, imbalanced, or discriminatory outcomes.
Bio Morphosis	An art project that explores AI-driven plant species fusion, raising ethical concerns about artificial manipulation, biodiversity conservation, and ecological impacts.
Black Box Society	A society where opaque systems govern information, decisions, and algorithms, hiding the inner workings of complex data-driven technologies.
Black swan	A black swan event is an unpredictable, rare occurrence with major consequences that can challenge AI systems, often disrupting standard risk assessment and preparedness strategies.
Bloomberg Cyborg	Generative AI that significantly transforms Bloomberg Media by automating data extraction and real-time news generation, dramatically increasing the speed and accuracy of financial journalism while promoting deeper human-AI collaboration in content creation.
Bot	An automated software application—for example, a chatbot or web crawler—that performs tasks or interactions, often simulating human behavior or conversation, and can operate independently or respond to specific commands across various platforms.

Bot-engagement	This concept refers to the interaction between humans and automated systems, such as AI-driven bots, in areas such as customer service, social media, and automated processes, optimizing efficiency and user experiences.
Business-driven morality	Ethical decisions and practices that are shaped primarily by corporate goals and profit motives—financial gain is prioritized over broader societal values.
Carbon capitalism	An economic system that incorporates the cost of carbon emissions into market mechanisms, aiming to reduce greenhouse gases through financial incentives and penalties.
Catalyst of transformative change	Generative AI drives change by transforming content creation, design, and fraud detection, advancing industries with automation, efficiency, scalability, and personalized solutions for varied needs.
Charismatic Project	AI-driven storytelling in media, supporting diverse creators in using generative AI to enhance film and television content creation and democratize technology.
Chinese room argument	A thought experiment from philosopher John Searle that challenges the notion of a computer's ability to understand or possess "true" intelligence merely by manipulating symbols.
Climate literacy	Understanding climate science and its effects, enabling informed decisions, with AI tools enhancing public awareness and promoting sustainable practices.
Co-intelligence	AI augments human abilities by assisting in decision-making, content creation, problem-solving, collaborating, and providing real-time suggestions while automating complex tasks.
Colonization	The exploitation of user data from marginalized communities to create profit-driven algorithms, reflecting historical patterns of resource exploitation and cultural erasure.
Commercialization and refeudalization of public discourse	AI-driven, profit-focused content prioritizes entertainment over meaningful debate, reinforces elite control of narratives, and limits public participation in rational, democratic discussions.
Complex narrative	Stories or information structures characterized by intricate plots, multiple characters, or layered themes, requiring significant engagement from the audience to understand.

Computational propaganda	The use of algorithms, automation, and AI by governments, organizations, or individuals to manipulate public opinion, spread mis/disinformation, or influence political outcomes.
Conspiracy theories	In generative AI, conspiracy theories are mainly driven by the rapid growth of synthetic content that reinforces harmful narratives. They thrive in echo chambers and ultimately distort public perception and increase mis/disinformation.
Constitutional AI	AI systems designed with ethical rules to ensure transparency, fairness, accountability, and alignment with human values in decision-making and actions.
Content creation processes	Generative AI transforms content creation by automating text, image, audio, and video generation, enabling real-time personalization and scaling creative output while minimizing manual effort in various industries.
Content mining	Extracting and processing large content-based datasets to train AI models and monetize these datasets by improving AI capabilities, often without direct remuneration to original content creators.
Contextual adaption	Adjusting systems, behaviors, or technologies to fit specific environmental, cultural, or situational contexts, increasing relevance and effectiveness by tailoring responses to unique needs.
Convergence	The merging of distinct technologies and industries enhances content creation, automates production, and personalizes experiences, driving innovation and creating new opportunities across sectors such as news, entertainment, and media.
Co-regulation	Governance model that combines industry self-regulation with government oversight, fostering shared responsibility and collaboration to address AI's rapid advancements and widespread applications across industries.
Creativity	In AI and GenAI context, it refers to generating novel, meaningful content by recognizing patterns, recombining data, and simulating innovation.
Creator economy	An economic system where creators use AI tools and platforms to generate content, products, or services, leading to new forms of creativity and monetization.

Crowdsourced content moderation	Enlisting a large group of online users or volunteers to monitor, review, and regulate user-generated content on digital platforms.
Crucible	A severe trial or test situation where different elements interact, creating something new or the emergence of fundamental changes.
Customization	Creating and delivering personalized content tailored to user preferences and behavior, amplifying engagement.
Dangerous determinism	The belief that all events, including human actions, are determined by external causes, undermining free will and moral responsibility.
Dangerous optimism	An overly optimistic and uncritical belief in the potential of technology, ignoring possible risks or adverse outcomes.
Data altruism	Willingly sharing data for the common good, or contributing to societal, scientific, or humanitarian goals without personal gain.
Data and AI monopolization	The dominance of a company that controls vast data resources and AI technologies, limiting competition by monopolizing the essential datasets and the advanced algorithms needed for innovation.
Data hunger	The growing demand for extensive data collection and analysis fueled by AI, IoT, and personalized services raises significant challenges related to privacy, storage, and energy consumption.
Data-intensive capitalism	An economic system where data collection, analysis, and exploitation are central to generating profit and gaining competitive advantages.
Data lakehouse	A data management architecture that merges the capabilities of a data lake (which stores vast amounts of raw data) with those of a data warehouse, enabling structured and queryable data.
Data School on Ethical AI, Utrecht University	The program designs tools for policymakers and civil servants to address ethical challenges of generative AI in media, ensuring responsible AI use, transparency, and accountability to protect public trust and human rights.
Data sovereignty	The principle that data is governed by the laws and regulations of the country where it is collected, ensuring national control over access, storage, and usage.

Datafication	Transforming social actions and processes into quantifiable data, allowing for analysis and optimization in various fields, from business to social sciences.
Datafied society	A social framework where personal information and behaviors are continuously collected, analyzed, and utilized to influence decision-making, drive business strategies, and shape public policies.
Dataset construction	Creating balanced datasets that ensure scale and quality while avoiding bias; they should reflect creators' judgments, for fairness and accountability in AI systems.
Dead End Gallery	The first gallery to exhibit works challenging traditional artistry, sparking debates about the value and authenticity of AI-generated art.
Decolonization	Deconstructing colonial ideologies, structures, and legacies, and valuing and revitalizing indigenous knowledge and practices.
DeepDream	A computer vision program that uses a convolutional neural network to find and enhance image patterns, creating dream-like, surrealistic artworks.
Deepfakes	The creation of hyperrealistic video or audio recordings using advanced AI techniques—i.e. replacing one person's likeness or voice with another's.
Deep mediatization	Integrating media into all aspects of society—transforming social, cultural, and economic processes through digital technologies, algorithms, and data-driven systems that shape daily life and communication.
Deep tech	Advanced technologies grounded in scientific breakthroughs, including AI, quantum computing, and biotechnology, which drive transformative innovations to tackle complex societal challenges while potentially causing long-term disruptions in various industries.
Degenerated generative AI	The deterioration of performance and output quality over time due to outdated data, reinforcing biases, or lack of updates and maintenance.
Democratic values	Principles that uphold democracy—including freedom of speech, equality, justice, the participation of citizens in political life—and their implications for AI governance.

Design discovery	Generative AI in design discovery autonomously generates and optimizes concepts, accelerating prototyping and fostering creative innovation in industries such as architecture, product development, and fashion.
Digital transformation	Integrating digital technology across all aspects of a business or society, fundamentally reshaping operations and optimizing value delivery to customers or citizens.
Disinformation	Intentionally disseminating false information to mislead or manipulate public opinion, often for political, social, or economic gain, undermining trust and informed decision-making.
Disruptive technology	Innovations that significantly alter the way businesses, industries, or markets operate, often displacing established technologies or practices.
Double-edged sword of AI innovation	While AI can bring significant benefits and advancements, it poses risks and challenges, such as ethical dilemmas, job displacement, and security threats.
Dual-use dilemma	The ethical challenge arising when scientific and technological advancements can be applied for both beneficial purposes and harmful outcomes, complicating responsible governance and oversight.
Dynamic human-machine interaction	An adaptive, real-time interaction model where machines employ AI to learn, personalize, and evolve responses to user behavior and shifting contexts.
Echo chambers	In generative technology, echo chambers are environments where algorithms amplify content that aligns with users' beliefs, restricting diverse perspectives and fostering mis/disinformation, thereby deepening societal divisions.
Edge computing	Processing data near its source at the network's edge, reducing latency, facilitating real-time processing, and improving efficiency by enabling local task management.
Education policy	The framework that directs the development, implementation, and regulation of educational systems and practices.
Elusive possibilities	Potential outcomes or opportunities that are difficult to achieve or predict due to their uncertain or speculative nature.

Emotion AI	Artificial intelligence systems designed to recognize, interpret, process, and simulate human emotions, often used in customer service, entertainment, and health care.
Emotional dependency	A psychological condition where an individual's emotional well-being is heavily reliant on the support, attention, or approval of other human or synthetic entities.
Entangled AI	A complex foresight of artificial intelligence that is deeply interconnected with humans, society, and nature, promoting a dynamic balance among social well-being, justice, and environmental resilience.
Ethical business models	Business strategies that prioritize ethical principles such as fairness, transparency, sustainability, and social responsibility in their operations and decision-making processes.
Ethical deliberation	The thoughtful consideration of moral principles and values when making decisions, especially in complex or ambiguous situations, with specific concerns about algorithm-driven decision-making.
Ethical scaling of moderation	The consistent application of moral principles in AI-driven content moderation across large platforms to ensure fairness, accountability, and user protection.
Ethics washing	The superficial adoption of ethical frameworks or guidelines to create a false image of responsibility without genuinely addressing deeper societal issues.
European Digital Media Observatory (EDMO)	A European Union initiative combating disinformation through collaboration among fact-checkers, researchers, and media organizations, focusing on media literacy and the effects of generative AI-driven content.
Explainable AI	An approach that enhances transparency in AI systems by clarifying how models make decisions, promoting trust and accountability through techniques that make AI processes understandable to users.
Extended experience	Enhanced or augmented experiences provided by technology (such as virtual reality or immersive multimedia) that go beyond traditional interactions.
Face Cages	An artistic, conceptual representation by Zach Blas that critiques surveillance, privacy invasion, and the

	loss of individuality, using physical or synthetic structures that resemble cages over faces.
Fairness	The quality of making judgments free from discrimination, bias, or favoritism, ensuring equal treatment and opportunities for all.
Fair use	A legal principle that allows limited use of copyrighted material without permission, for purposes such as criticism, education, or research, considering factors like purpose, amount, and market impact.
Fake news factories	Organizations or entities that systematically disseminate false information or propaganda, often for political or financial gain.
FBI's AI-Detected GenAI Fraud Case	This streaming scam involved the use of generative AI to create fake songs. Bots were used to inflate streams and collect fraudulent royalties, which was detected by advanced AI algorithms monitoring irregular activity.
Feature engineering	The process of selecting, modifying, or creating new features from raw data to improve the performance of machine learning models.
Filter bubble	An online environment where algorithms personalize content to reinforce a user's beliefs and preferences, limiting exposure to diverse perspectives.
Fluid ethics	The adaptability of ethical frameworks in response to AI technologies' complexities and unpredictable nature. It emphasizes the need for evolving standards in decision-making.
Fluid GenAI ethics	Explores the ethical challenges of generative AI outputs and frameworks tailored to real-time data and synthetic outputs.
Fluid nature of generative AI	Generative AI's ability to adapt in real time to contextual changes, with learning models that evolve continuously based on new data, resulting in a dynamic and responsive system.
Foundational models	Large-scale, pre-trained AI models that serve as a base for fine-tuning across various tasks, leveraging massive datasets and general representations.
Four waves of AI	The first was handcrafted knowledge based on predefined rules; the second was statistical learning that leverages data patterns; the third was contextual adaptation facilitating nuanced interaction; and the fourth is generative AI, which produces synthetic

	content that challenges human agency and human perception of truth or facts.
Fraud detection	Generative AI improves fraud detection by spotting patterns and anomalies in large datasets, enabling financial institutions and e-commerce platforms to predict and prevent fraud more accurately and efficiently.
Freemium model	A business strategy offering free basic services or products while charging for premium features or additional functionalities.
Future AI media	Encompassing advanced media production and services powered by recognition-behavioral-transaction technologies and sophisticated analytics.
Future of generative AI	The concept promises an evolution in how we create or conceptualize something new, produce knowledge or services, and interact with content, blending synthetic and human-centered approaches.
Future prediction	Using AI and algorithms to forecast events or trends based on historical data and patterns.
Generative Adversarial Networks (GANs)	A machine learning framework where two neural networks compete to create and evaluate realistic synthetic data, a generator, and a discriminator.
Generative AI	Generative AI (GenAI) is a technology that creates synthetic outputs—such as content, designs, and models—by learning patterns from input data and prompts, fundamentally transforming creativity, business processes, and societal structures.
Generative AI brokers	Intermediaries between developers, policymakers, and the public, ensuring ethical generative AI deployment by managing societal risks.
Generative AI detection	Techniques or systems used to identify outputs that AI has generated, distinguishing them from human-created content.
Generative AI ethics	An interplay of business-driven morality and pro-ethical generative AI design in the society-human-machine interaction.
Generative AI for good	The application of generative AI technologies to address social and environmental challenges, producing innovative outputs that promote social responsibility, community well-being, and sustainable development.
Generative AI literacy	Integrating technology education from early childhood through universities, involving businesses,

	educators, and parents to foster critical, ethical technology use across all generations.
Generative AI media	AI-driven content creation using machine learning models, vast datasets, and statistical techniques to generate synthetic media that mimic human-made outputs to transform cultural conversations, public discourse, and immersive experiences.
Generative AI policy	Frameworks governing generative AI technologies, focusing on intellectual property, mis/disinformation, social impact, and data privacy while addressing AI's ecological footprint and promoting sustainable practices.
Generative AI value chain	The processes involved in training, deploying, and monetizing generative AI, from data collection and model development to content creation, customization, and feedback-driven improvement.
Generative colonization	The replication and expansion of biased narratives in synthetic content, reinforcing existing societal power dynamics through algorithmically generated outputs.
Generative Explainable AI (GenXAI)	A framework that enables stakeholders to comprehend and verify the processes by which generative AI models create synthetic outputs, ensuring transparency in decision-making, fostering user trust, and minimizing potential misuse.
Generative ghosts	Digital artifacts created from individuals' online activities and interactions, forming AI-generated personas or representations that remain after their physical departure, embodying their past digital engagement and identity.
Generative mediatization	This concept explores how AI algorithms transform media production, modification, and distribution while serving as disruptors and catalysts that influence public discourse, media practices, and societal expectations.
Generative morality	Integration of ethical principles and moral frameworks into generative AI systems, producing outputs that reflect societal values through dynamic interactions and creative processes.
Generative Responsible AI (GenRAI)	A framework that guides generative AI systems to align with ethical standards, ensuring content creation upholds fairness, transparency, and accountability, reducing risks of bias, privacy violations, and mis/disinformation.

Generative search	An AI-driven search mechanism that creates or generates new content or answers in response to queries rather than simply retrieving existing information.
Genetic algorithms	Optimization techniques inspired by evolution that use natural selection, crossover, and mutation to solve complex problems through the evolution of solutions over generations.
Geopolitics of AI	Strategies for AI development and control by various nations and regions that influence global power dynamics, security, and competitive advantage through governance, data sovereignty, and technological leadership.
Glass box AI	Transparent, interpretable AI for human-in-the-loop and trust.
Governing AI	Development and implementation of regulatory frameworks, policies, and ethical guidelines that ensure the development, deployment, and oversight of artificial intelligence technologies, ideally aligned with societal values, legal standards, and equitable use across industries and sectors.
Half-life of knowledge	The time it takes for half of the knowledge in a specific field to become obsolete due to technological advancements, innovations, or discoveries.
Hallucination	Generation of content that appears coherent but requires more precise understanding, often leading to contextually inaccurate or nonsensical outputs.
Hereafter	An art project exploring memory, loss, and technology by creating generative ghosts from personal digital footprints, prompting reflection on synthetic legacy and human immortality.
Hologram	A three-dimensional image created with photographic projection, often used in entertainment, presentations, and as a method of communication.
Human-centered AI	AI development that prioritizes human values, needs, and ethical considerations, ensuring that AI systems enhance human capabilities and well-being.
Human rights	Fundamental rights and freedoms to which all individuals are entitled, including the right to life, liberty, and security.
Human workforce	The segment of the population actively engaged in or available for employment across various labor markets and industries.
Hyperpersonalization	Using real-time data, behavioral insights, and AI to deliver highly individualized experiences, dynamically

	tailoring products, services, and communications to meet specific customer needs across industries.
Hyperproduction	Large-scale automation of content creation through AI, blurring the line between real and synthetic content and transforming how media is produced, consumed, and commodified.
Hyperreality	An augmented or enhanced reality created by combining real and virtual elements, often blurring the line between what is real and what is simulated.
Imagine	An art project that uses AI and fMRI scans to generate images from human thoughts. It explores the fusion of human cognition and AI, raising ethical concerns about privacy and creativity.
Inclusivity	Diverse and equitable participation in the innovation process, ensuring that technological advancements benefit all segments of society and promote fairness and shared progress.
In Event of Moon Disaster	A deepfake project that imagines President Nixon announcing an unsuccessful Apollo 11 landing, examining AI's ethical implications for truth, memory, and the spread of mis/disinformation.
Infinite media	Digital technologies and the Internet make limitless media content possible, leading to overwhelming amounts of information.
Innovation imperatives	Principles of inclusivity, sustainability, adaptability, collaboration, and ethical responsibility that guide technological advancements to meet diverse societal needs and promote responsible development.
Institutionalized activism in human rights	Involvement of minorities, subcultures, and Indigenous people in partnerships with businesses, academia, and policymakers, ensuring their representation in responsible AI and equity-focused policies.
Integrative ethics	Multifaceted AI ethics framework that harmonizes top-down regulations with grassroots moral engagement, blending diverse ethical theories.
Intellectual property rights	Legal rights that protect the creations of the mind, such as inventions, literary and artistic works, designs, symbols, names, and content used in commerce.
Intelligent agent	An autonomous entity capable of perceiving its environment, processing information, and making decisions to achieve specific goals, often utilizing machine learning and artificial intelligence techniques.

ISO standards for generative AI	A set of internationally recognized guidelines ensuring privacy, security, risk management, and ethical compliance in generative AI systems to uphold transparency, accountability, and explainability in AI governance.
Kantianism	A moral philosophy emphasizing duties and respect for individual rights and autonomy. Its application ensures that AI systems treat users as ends in themselves, prioritizing transparency, truthfulness, and fairness over outcomes.
Life-centered design	An approach to design that prioritizes the well-being and sustainability of all life forms and the environment in developing technologies.
Machine behavior	The study of how AI-driven systems act, make decisions, and interact with their environment, influencing societal transformation by automating tasks, shaping behaviors, and impacting human-machine dynamics across various fields.
Manipulation	Skillfully controlling or influencing media content to shape perceptions and achieve a desired outcome through deliberate tactics and strategic messaging.
Measurable probabilities	Quantifiable likelihoods of various outcomes, often expressed as percentages or ratios.
Media	A catalyst that explores and shapes the intersections of technology, social relations, economic models, political interests, and aesthetic expressions, influencing our perceptions and engagement with the world across various technologies.
Media credibility	The trustworthiness of AI-generated content emphasizes accuracy, transparency, and ethical standards, especially in generative or synthetic media production.
Media's portrayal	How artificial intelligence is represented and discussed in various media formats, influencing public understanding and attitudes.
Mediatization	The process by which media becomes a central force in shaping socio-cultural norms and values.
Memory effect	AI-generated content subtly alters personal memories and perceptions, blurring the line between authentic experiences and artificial creations and undermining societal trust in shared histories.
Metamedia	Integrating diverse media forms through computing platforms fosters active engagement and

	customization, surpassing traditional multimedia to create dynamic, interactive experiences.
Metareasoning	The process of reasoning about one's cognitive processes and deciding how to think, or solve problems.
Metaverse	A collective virtual shared space created by converging AI-driven, virtually enhanced physical and digital reality.
Mind control	Potentially manipulating users' thoughts and behaviors through immersive technologies—that is, directly influencing perceptions and actions by integrating AI, neural interfaces, and real-time generated content.
Mirror Test	An art project exploring AI's ability to reflect human knowledge and language through behavioral psychology, prompting reflection on GenAI's evolving capabilities and potential self-awareness.
Misinformation	Unintentionally sharing incorrect or misleading information without the intent to deceive.
Misuse risks	The potential dangers associated with the improper or unethical use of technology, information, or resources.
MIT Consortium's Project in GenAI's Environmental Impact	This initiative tackles generative AI's environmental impact, advocating for sustainable development by balancing AI productivity with energy consumption and aiming for net-zero emissions.
Moral compass	The set of ethical guidelines and principles that govern the development, deployment, and operation of AI systems, ensuring they make decisions that align with human values.
Neuralink	A technology company focused on developing implantable brain-machine interfaces.
Neuromorphic computing	A type of computing that mimics the human brain's neural structure and operation.
New-smart-AI media evolution	The evolution of media platforms to enhance content creation, distribution, and interaction, moving from new media to smart services and ultimately advancing into AI-enhanced media.
NewsTech	Integrating AI and algorithms into journalism, transforming traditional news production and dissemination through enhanced content creation and distribution methods.
Open-source ecosystem	A collaborative software development framework that fosters innovation but faces challenges in quality

	control, security, and managing decentralized contributions, requiring a balance between openness, governance, and stability.
OSPOs under the United Nations Global Digital Cooperation	Open Source Program Offices promote scalable, transparent digital technologies for global collaboration, address ethical challenges, and support sustainable digital-AI transformation.
Ouroboros	The self-consuming cycle in which generative AI processes its synthetic outputs, risking content degradation, narrowing perspectives, and amplifying uncertainties within the system.
Perception	The process of interpreting sensory information to represent and understand the environment around us.
Personalization	Methods to tailor products, services, or content to individual users or customers based on their preferences, behaviors, or characteristics.
Platform and AI governance	Regulatory frameworks and policies that ensure accountability and ethical use of digital platforms and AI, addressing data privacy, bias, transparency, and environmental impact.
Platformization	The process by which digital platforms increasingly dominate and shape social, economic, and cultural activities, integrating various services and mediating interactions through centralized infrastructures.
Platform monopoly	A single digital platform that dominates a market, restricting competition and controlling user access, data, and services, influencing market dynamics and innovation.
Poisoned training data	Data that has been intentionally altered or corrupted to compromise the performance of machine learning models.
Polanyi's paradox	The observation that we know more than we can tell refers to the tacit knowledge that is difficult to transfer to others or AI.
Political bots	These bots manipulate public opinion by amplifying specific messages or spreading disinformation through social media.
Political economy	How political institutions, processes, and structures influence economic policies, outcomes, and relationships, examining the interplay between politics and economics in shaping societal well-being.

Political economy of generative AI	The concept explores synthetic data's role in redefining data ownership and value, challenging traditional economic models and expanding AI-driven capitalism.
Posthumanism	A philosophical approach that explores the evolution of human identity, focusing on integrating technology and the dissolution of traditional boundaries between humans, machines, and the environment, envisioning a future beyond human-centric perspectives.
Post-public concept	Public discourse becomes fragmented as authentic and manipulated content blends, driven by human and AI agents, disrupting traditional democratic engagement.
Post-truth	A socio-cultural and political environment where objective facts are less influential in shaping public opinion than appeals to emotion and personal beliefs, allowing mis/disinformation to spread.
Privacy	The right to keep one's personal information and activities confidential and free from unauthorized surveillance or disclosure.
Proactive strategy	Anticipating future challenges and opportunities, enabling organizations to take preemptive actions rather than merely reacting to events.
Process journalism	A journalistic approach that emphasizes news publication in its developing phases, allowing for updates and corrections as more information becomes available.
Pro-ethical design	The deliberate development of AI systems that prioritize ethical values, including fairness, transparency, and societal well-being, above profit-driven objectives.
Prompt	In generative AI, a prompt is a user-provided input or instruction that directs the AI to generate specific content, such as text, images, media, music, content, code, or even designs and models, based on the provided guidance.
Prompt engineering	Designing and refining prompts to effectively guide generative AI models, optimizing their outputs.
Quantum	Relating to the most minor possible discrete units of any physical property, often used in physics to describe particles and their behaviors.

Quantum computing	A transformative computing paradigm that leverages the principles of quantum mechanics, specifically using quantum bits (qubits) to perform computations.
Quantum encryption	A security tool using the principles of quantum mechanics, like superposition and entanglement, to secure data, ensuring that any interception of the encryption keys is immediately detectable.
Recommendation systems	Algorithms designed to suggest items to users based on their preferences and behaviors.
Regime of hope	A sociological concept referring to the optimistic belief in the progressive potential of a particular technology or innovation.
Reinforcement Learning from Human Feedback (RLHF)	A method where generative AI models improve performance by making decisions, receiving reward feedback, and learning from the outcomes.
Relational ethics	The concept focuses on the contextual and dynamic nature of interactions between individuals and systems, emphasizing that AI operates within social and cultural networks, impacting decision-making and public trust.
Religious AI	AI systems designed to interact with or interpret religious texts, doctrines, or practices, providing insights, teachings, or guidance based on those religious frameworks.
Responsible AI	Socio-cultural construction throughout the AI lifecycle, ensuring accountability, fairness, and transparency.
Reward system	Feedback-driven models, optimizing content based on rewards for successful outcomes.
Rights-based governance model	Collective data ownership and equitable participation in AI development, ensuring fair distribution of technological benefits and addressing risks through inclusive decision-making frameworks.
Risk	In AI, potential adverse outcomes include ethical dilemmas, privacy breaches, job displacement, biases in decision-making, and unintended consequences from autonomous systems.
Robot	An autonomous machine senses its surroundings, makes decisions, and performs tasks, either fixed or mobile, reprogrammable, and multipurpose.
Robot journalism	The automated production of news stories and reports by AI algorithms without direct human authorship.

Saint George on a Bike	AI and supercomputing are used to analyze and enrich European cultural heritage images, enhancing public access, understanding, and contextual accuracy through automated descriptions and crowd-sourced annotations.
Scalability	The system's capacity to expand efficiently to handle increased workloads—such as larger datasets or complex tasks—without sacrificing performance, quality, or speed.
Scaling	Expanding a system's capacity to handle larger datasets, more complex models, and broader applications, improving output quality, efficiency, and performance across industries.
Self-regulation in AI	Refers to the autonomous establishment and adherence to ethical standards by industries or organizations, balancing technological benefits with social risks and focusing on responsible use without external oversight.
Silent intelligence	Unobtrusive, automated processing and decision-making capabilities of AI systems, operating behind the scenes to enhance efficiency, personalization.
Social contract	An implicit agreement that establishes the expectations and responsibilities between AI systems, users, and society, focusing on ethical use, transparency, accountability, and protecting individual rights in the context of AI-generated content.
Societal-environmental sustainability	Strives for technological optimization while minimizing societal harm and ecological impact, balancing energy use for future generations.
Social generative AI responsibility	The ethical duty to create generative AI that ensures fairness, transparency, and inclusivity, to address issues such as authorship, mis/disinformation, and bias, with frameworks for accountability and societal impact.
Social good	Fostering collective well-being, transcending individual or corporate gain, and contributing to society's overall progress and sustainability.
Societal inequities	The unequal and unfair distribution of resources, opportunities, and rights among different societal groups.
Societal norms	Shared expectations and rules that guide societal behavior, often unwritten and learned through social interaction.

Societal scale of generative AI	The extensive impact of AI-generated content across media and further industries influences public perception, ethics, and policy while transforming cultural norms and economic systems worldwide.
Society-in-the-loop	A framework where societal feedback influences AI development, aiming to ensure alignment with social structures or modification of them.
Socio-informatics	Multidisciplinary study of the interaction between social structures, human behavior, and information technologies, focusing on the societal implications of digital and AI systems for culture, communication, and the co-evolution of technology with humanity.
Sociopath generative AI	When a generative AI system produces speculative or fictional outputs that lack empathy and moral considerations.
Socio-technical errors	Mistakes or failures arising from complex interactions between human (social) and technological systems.
Socio-technical systems	Integrated systems that involve a complex interaction between humans (social aspects) and technology (technical aspects).
Spatial journalism	Using AI and spatial data by *The New York Times* R&D, this project creates immersive, location-specific news narratives, addressing privacy, data ethics, and accuracy in AI-driven journalism.
Stable world principle	A principle suggesting that a system (such as an AI system) must operate within a stable and predictable framework that aligns with human values and societal norms to be ethical and practical.
Static human-machine interaction	A fixed, rule-based approach to human-computer interaction where machine responses are predetermined, lacking adaptability or learning, resulting in limited responsiveness to varied user inputs.
Stochastic parrots	A term that describes AI models that mimic human language by randomly generating text based on patterns learned from large datasets without understanding or intention.
Strategic foresight	A systematic process of anticipating future trends, risks, and opportunities to guide decision-making and governance, enabling the implementation of long-term strategies and effective uncertainty management.
Subscription model	Widely used business model offering tools, platforms, and services through recurring fees and continuous access.

Super App	A multifunctional platform that combines various services into a single application, utilizing AI to enhance user experience through personalization, convenience, and automation.
Surveillance	The close observation or monitoring of individuals or groups, often to manage, influence, or protect them.
Sustainability Film	A film by Beko that utilizes Generative Adversarial Networks (GANs) to blend AI-generated visuals with live-action footage. It emphasizes sustainability and ethical considerations while demonstrating AI's potential in creative media.
Synthetic emotion	Simulation or generation of emotional responses by artificial systems that mimic human emotional expression without experiencing genuine feelings.
Synthetic influencers	AI-driven digital characters capable of autonomously generating content and interactions. They adapt based on data and audience engagement and offer more independent behavior than virtual influencers.
Talent economy	Economic model where growth is driven by skilled individuals, with policies focusing on developing, attracting, and retaining talent to foster innovation and address evolving industry demands.
Technological determinism	The theory posits that technological development is the primary driver of societal change, shaping human behavior and cultural norms more significantly than social, economic, or political factors.
Techno-narcissism	An excessive fixation on technology as the ultimate solution, often accompanied by disregard for human values, social relationships, and ethical considerations—in favor of technological advancements.
Technosolutionism	The belief that complex social and environmental issues can be solved solely through technological innovations often without looking at societal and ethical implications.
The Next Rembrandt	A deep learning art project that used data from Rembrandt's paintings to create a new artwork mimicking his style. It explored the intersection of art, technology, and data analysis.
TPESTRE Model	An analytical framework that includes Technological, Political, Economic, Social, Technological, Regulatory, and Environmental factors used for strategic planning.

Transhumanism	A movement that advocates using advanced technologies to enhance human physical, cognitive, and emotional abilities, aiming to transcend biological limitations, including aging and disease, and improve human well-being.
Transparency	A principle ensuring AI systems are understandable and accessible, allowing stakeholders to trust and scrutinize outcomes.
Trusted AI	AI systems designed to be reliable, safe, and worthy of confidence, often through ethical practices and fairness.
Trust Paradox	The phenomenon where individuals cannot trust information from sources they do not understand, with their lack of trust preventing them from seeking to understand those sources.
Turing Test	A test of a machine's ability to exhibit intelligent behavior equivalent to, or indistinguishable from, that of a human, proposed by Alan Turing.
Turing Trap	A concept warning against developing human-like AI that automates rather than augments labor, risking concentrated wealth and power, which could hinder equitable productivity gains and limit societal improvement.
Uncertainties	The unpredictable risks and complexities of integrating advanced technologies, which require structured management, assessment, and reflection for responsible development.
Uncertainty management	Addresses unpredictability by employing strategies to assess risks, manage incomplete data, and design adaptable models, ensuring robust, reliable decision-making in complex or ambiguous situations.
Unsupervised	A digital artwork that uses GenAI to transform archival images and metadata into fluid visuals, redefining audience engagement with art and marking a breakthrough in AI-driven artistic expression.
User engagement	The extent to which users interact with and are involved in a platform, product, or service.
Utilitarianism	Maximizing overall benefits and minimizing harm by ensuring AI systems produce outcomes that provide the greatest good for the most significant number of people, even if some ethical risks are taken.
Utopia vs. dystopia	The contrasting visions of ideal perfection and catastrophic societal decline are often explored in the context of technological advancement.

Victimization in the Metaverse by Deepfake	An art project exploring harm and exploitation within virtual worlds and AI-driven platforms, provoking reflection on safety, privacy, and ethical concerns.
Virtual influencers	AI-driven, computer-generated characters managed by humans, crafted to engage online audiences through curated content and designed to mimic human behavior, appearance, and personality traits.
Vulnerability	A flaw or weakness in a system that can be exploited to compromise its security, functionality, and reliability.
Watermarking	A technique used to embed a hidden marker or logo in a document or data to protect copyright, verify authenticity, or signal ownership.
Weapon of mass deception	A term that refers to tools or strategies designed to manipulate public opinion or spread mis/disinformation on a large scale.
Whataboutism	A rhetorical tactic that deflects criticism by raising unrelated issues or accusations to shift the focus away from the original topic, to avoid direct engagement.
What Models Make Worlds	The Ford Foundation Gallery's exploration of AI's societal impact, addressing justice, inclusivity, and transparency, and critiquing embedded biases in AI through themes like surveillance, data erasure, and discrimination.
Wild card	A highly improbable yet impactful event that can significantly alter trends or assumptions within a system.

REFERENCES

Aggarwal, S., Sinha, T., Kukreti, Y., & Shikhar, S. (2020). Media bias detection and bias short term impact assessment. *Array*, 6, 100025.

AI Action Summit (2025). International AI Safety Report. https://assets.publishing.service.gov.uk/media/679a0c48a77d250007d313ee/International_AI_Safety_Report_2025_accessible_f.pdf

Aitamurto, T., Aymerich-Franch, L., Saldivar, J., Kircos, C., Sadeghi, Y., & Sakshuwong, S. (2022). Examining augmented reality in journalism: Presence, knowledge gain, and perceived visual authenticity. *New Media & Society*, 24(6), 1281–1302.

Ali Abbasi, G., Abdul Rahim, N. F., Wu, H., Iranmanesh, M., & Keong, B. N. C. (2022). Determinants of SME's social media marketing adoption: Competitive industry as a moderator. *Sage Open*, 12(1), doi:21582440211067220.

Ananny, M. (2024a). Making generative artificial intelligence a public problem: Seeing publics and sociotechnical problem-making in three scenes of AI failure. *Javnost –The Public*, 31(1), 89–105.

Ananny, M. (2024b,July 16). Recursive press freedom as the capacity to control and learn from mistakes. *Knight First Amendment Institute at Columbia Institute*. https://knightcolumbia.org/content/recursive-press-freedom-as-the-capacity-to-control-and-learn-from-mistakes.

Ananny, M. & Finn, M. (2020). Anticipatory news infrastructures: Seeing journalism's expectations of future publics in its sociotechnical systems. *New Media & Society*, 22(9), 1600–1618.

Ananny, M. & Karr, J. (2023, September 27). Press freedom means controlling the language of AI. *Nieman Lab, Harvard*. https://www.niemanlab.org/2023/09/press-freedom-means-controlling-the-language-of-ai/.

Apablaza-Campos, A., Tinjaca, J. A. W., & Salaverria, R. (2024). Generative artificial intelligence for journalistic content in Ibero-America: Perceptions, challenges and regional projections. *BiD: textos universitaris de biblioteconomia i documentació*, 52.

Appelganc, K., Rieger, T., Roesler, E., & Manzey, D. (2022). How much reliability is enough? A context-specific view on human interaction with (artificial) agents from different perspectives. *Journal of Cognitive Engineering and Decision Making*, 16(4), 207–221.

Araujo, T., Helberger, N., Kruikemeier, S., and De Vreese, C. H. (2020). In AI we trust? Perceptions about automated decision-making by artificial intelligence. *AI & Society*, 35, 611–623.

Arguedas, A. R. & Simon, F. M. (2023, July). Automating democracy: Generative AI, journalism, and the future of democracy. *Oxford University Institute for AI Ethics*. https://www.oii.ox.ac.uk/news-events/reports/automating-democracy-generative-ai-journalism-and-the-future-of-democracy/.

Arifin, A. A. & Lennerfors, T. T. (2021). Ethical aspects of voice assistants: A critical discourse analysis of Indonesian media texts. *Journal of Information, Communication and Ethics in Society*, 20(1), 18–36.

Artificial Intelligence Center of Excellence. (2023). *AI Guide for Government*. https://coe.gsa.gov/ai-guide-for-government/.

Ashrafian, H. (2015). Artificial intelligence and robot responsibilities: Innovating beyond rights. *Science and Engineering Ethics*, 21, 317–326.

Ashworth, M. (2023, May). *Fascinated and Frightened: How are Communications Professionals Viewing the AI Opportunity Ahead?* WE Communications. https/www.we-worldwide.com/media/i5sjuvsa/fascinated-and-frightened-ai-in-comms_we-annenberg-2023.pdf.

Aslett, K., Sanderson, Z., Godel, W., Persily, N., Nagler, J., & Tucker, J. A. (2024). Online searches to evaluate misinformation can increase its perceived veracity. *Nature* 625, 548–556.

Aue, J. (2018, July 11). The human-to-machine communication model. *Medium*. https://medium.com/design-ibm/the-human-to-machine-communication-model-d630ec54a12e.

Balaji, N. (2023, June 7) ChatGPT may create deadly polymorphic malware that evades EDR. *Cyber Security News*. https://cybersecuritynews.com/chatgt-may-create-deadly-polymorphic-malware-that-evades-edr/.

Barassi, V. (2020). Datafied times: Surveillance capitalism, data technologies and the social construction of time in family life. *New Media & Society*, 22(9), 1545–1560.

Barcelona Supercomputing Center. (2022). Saint George on a Bike. *Heritage Research Hub*. https://www.heritageresearch-hub.eu/project/saint-george-on-a-bike/.

Bareis, J. & Katzenbach, C. (2022). Talking AI into being: The narratives and imaginaries of national AI strategies and their performative politics. *Science, Technology, & Human Values*, 47(5), 855–881.

Batori, Z. (2024). *AI-generated photo-based images: Their ontological status and interpretation*. The Annual Conference of the European Society for Aesthetics, Naples, Italy, June 7.

Baxter, K. & Schlesinger, Y. (2023, June 6). Managing the risks of generative AI. *Harvard Business Review*. https://hbr.org/2023/06/managing-the-risks-of-generative-ai.

Bekker, A. (2019). 5 types of AI to propel your business head of data analytics department, *ScienceSoft, IT Consulting & Software Development*. https://www.scnsoft.com/blog/artificial-intelligence-types.

Beko. (2022). Behind the Scenes. *YouTube*. https://www.youtube.com/watch?v=TfYnxeO5BTg.

Bender, E. M., Gebru, T., McMillan-Major, A., & Shmitchell, S. (2021, March). On the dangers of stochastic parrots: Can language models be too big? *Proceedings of the 2021 ACM Conference on Fairness, Accountability, and Transparency* (pp. 610–623). New York: Association for Computing Machinery.

Bender, S. (2024). Generative-AI, the media industries, and the disappearance of human creative labour. *Media Practice and Education*, 1–18.

Bengio, Y. & Privitera, D. (2023, November 8). How we can have AI Progress without sacrificing safety or democracy. *Time*. https://time.com/collection/time100-voices/6325786/ai-progress-safety-democracy/.

Bevilacqua, M., Berente, N., Domin, H., Goehring, B., & Rossi, F. (2023). The return on investment in AI Ethics: A holistic framework. *arXiv*. doi:2309.13057.

Biamonte, J., Wittek, P., Pancotti, N., Rebentrost, P., Wiebe, N. & Lloyd, S. (2017). Quantum machine learning. *Nature*, 549, 195–202.

Biggins, M. (2017, September 12). How the world is preparing for the AI apocalypse. *Medium*. https://medium.com/s/ai-dirty-little-secret/how-to-prepare-for-the-ai-apocalypse-80fb815a9166.

Bingley, W. J., Wittek, P., Pancotti, N., Rebentrost, P., Wiebe, N. & Lloyd, S. (2023). Where is the human in human-centered AI? Insights from developer priorities and user experiences. *Computers in Human Behavior*, 141, 107617.

Blas, Z. (2016). Face Cages. *Zach Blas*. https://zachblas.info/works/face-cages/.

Bodo, B. (2021). Mediated trust: A theoretical framework to address the trustworthiness of technological trust mediators. *New Media & Society*, 23(9), 2668–2690.

Bogost, I. (2022, December 7). ChatGPT is dumber than you think. *The Atlantic*. https://www.theatlantic.com/technology/archive/2022/12/chatgpt-openai-artificial-intelligence-writing-ethics/672386/.

Bol, N., Strycharz, J., Helberger, N., van de Velde, B., & de Vreese, C. H. (2020). Vulnerability in a tracked society: Combining tracking and survey data to understand who gets targeted with what content. *New Media & Society*, 22(11), 1996–2017.

Borup, M., Brown, N., Konrad, K. & Van Lente, H. (2006). The sociology of expectations in science and technology. *Technology Analysis & Strategic Management*, 18(3–4), 285–298.

Boston Global Forum. (2020, September 9). Social contract for the AI age. *Boston Global Forum*. https://bostonglobalforum.org/initiative/aiws-and-the-age-of-global-enlightenment/social-contract-for-the-ai-age/.

Breazu, P. & Katsos, N. (2024). ChatGPT-4 as a journalist: Whose perspectives is it reproducing?. *Discourse & Society*, 35(6), 687–707.

Brevini, B. (2021). Carbon capitalism, communication, and artificial intelligence: Placing the climate emergency center stage. In López, A. *et al.* (eds), *The Routledge Handbook of Ecomedia Studies* (pp. 171–178). Oxford and New York: Routledge.

Brevini, B. & Pasquale, F. (2020). Revisiting the Black Box Society by rethinking the political economy of big data. *Big Data & Society*, 7(2).

Brinkhof, T. (2021, August 23). How to paint like Rembrandt, according to artificial intelligence. *Discover*. https://www.discovermagazine.com/technology/how-to-paint-like-rembrandt-according-to-artificial-intelligence.

Broussard, M., Diakopoulos, N., Guzman, A. L., Abebe, R., Dupagne, M., & Chuan, C. H. (2019). Artificial intelligence and journalism. *Journalism & Mass Communication Quarterly*, 96(3), 673–695.

Brown, S. (2021, June 16). The case for new social media business models. *MIT Management Sloan School*. https://mitsloan.mit.edu/ideas-made-to-matter/case-new-social-media-business-models.

Browning, J. & Lecun, Y. (2023, August 23). AI and the limits of language. *Noema*. https://www.noemamag.com/ai-and-the-limits-of-language/.

Bruell, A., Schechner, S., and Seetharaman, D. (2024, May 22). OpenAI, WSJ Owner News Corp Strike Content Deal Valued at Over $250 Million. *The Wall Street Journal*. https://www.wsj.com/business/media/openai-news-corp-strike-deal-23f186ba.

Brylska, K., Balcerzak, A., Gackowski, T., Kowalczyk, G., Łączynski, M., Mierzecka, A., Sztyber, M., Blak, A., Hykawy, J., Szydelski, J., Sarna, M., & Zakrzewska, J. (2024). *Poles on Artificial Intelligence: CATI Survey Report by LBM UW*. Laboratorium Badań Medioznawczych UW. https://www.lbm.uw.edu.pl/images/Raport_AI_ENG2.pdf.

Brynjolfsson, E. (2022). The Turing Trap: The promise & peril of human-like artificial intelligence. *Daedalus*, 151(2), 272–287. https://digitaleconomy.stanford.edu/news/the-turing-trap-the-promise-peril-of-human-like-artificial-intelligence.

Brynjolfsson, E. & McAfee, A. (2014). *The Second Machine Age: Work, Progress, and Prosperity in a Time of Brilliant Technologies*. New York: WW Norton & Company.

Burke, G. & Schellmann, H. (2024, 26 October). Researchers say an AI-powered transcription tool used in hospitals invents things no one ever said. *Associated Press*. https://apnews.com/article/ai-artificial-intelligence-health-business-90020cdf5fa16c79ca2e5b6c4c9bbb14.

Calice, M. N., Bao, L., Freiling, I., Howell, E., Xenos, M. A., Yang, S., Brossard, D., Newman, T. P., & Scheufele, D. A. (2023). Polarized platforms? How partisanship shapes perceptions of "algorithmic news bias". *New Media & Society*, 25(11), 2833–2854.

California Legislature. (2023/AB331). California AI Transparency Act (SB-942). *California Legislative Information*. https://leginfo.legislature.ca.gov/faces/billNavClient.xhtml?bill_id=202320240SB942.

Campos-Freire, F., Rúas-Araújo, J., López-García, X., and Martínez-Fernández, V. (2016). The impact of social media on journalism. *Profesional de la Informacion*, 25 (3), 449–457.

Carroll, M. (2023). Deutsche Telekom boss laments decline of Europe. *Mobile World Live*. https://mobileinsights.mobileworldlive.com/mwc-barcelona-2023-dailies/deutsche-telekom-boss-laments-decline-of-europe.

Caselles-Dupre, H., Mellerio, C., H'erent, P., Lopez-Persem, A., B'eranger, B., Soularue, M., Fautrel, P., Vernier, G., and Cord, M. (2024). Mind-to-Image: Projecting visual mental imagination of the brain from fMRI. *arXiv preprint*. doi:2404.05468.

Castells, M. (1996). *Rise of the Network Society: The Information Age: Economy, Society and Culture*. Oxford: Blackwell.

Castillo-Campos, M., Varona-Aramburu, D., & Becerra-Alonso, D. (2024). Artificial intelligence tools and bias in journalism-related content generation: Comparison between Chat GPT-3.5, GPT-4 and Bing. *Tripodos*, (55), 99–115.

Chaillou, S. (2024). *Bio Morphosis*. http://stanislaschaillou.com/biomorphosis.html.

Chadwick, A., Hall, N. A., & Vaccari, C. (2023). Misinformation rules!? Could "group rules" reduce misinformation in online personal messaging? *New Media & Society*, doi:14614448231172964.

Chan-Olmsted, S. M. (2019). A review of artificial intelligence adoptions in the media industry. *International Journal on Media Management*, 21(3–4), 193–215.

Charles, P. (2019, January 8). What can you do about the half-life of your professional knowledge? *Medium*. https://philch.medium.com/what-can-you-do-about-the-half-life-of-your-professional-knowledge-6ac3812ef329.

Chen, L., Zaharia, M., & Zou, J. (2023). How is ChatGPT's behavior changing over time? *Arxiv*. https://arxiv.org/abs/2307.09009.

Chesbrough, H. W. (2003). *Open Innovation: The New Imperative for Creating and Profiting from Technology*. Boston, MA: Harvard Business School Publishing Company.

Chhillar, D. & Aguilera, R. V. (2022). An eye for artificial intelligence: Insights into the governance of artificial intelligence and vision for future research. *Business & Society*, 61(5), 1197–1241.

Chin, C. (2021, February 2). Reduce noise, not cognitive biases. *Commoncog*. https://commoncog.com/reduce-noise-not-cognitive-biases/.

Christensen, C. M. (2015). *The Innovator's Dilemma: When New Technologies Cause Great Firms to Fail*. Boston, MA: Harvard Business Review Press.

Christodoulou, E. & Iordanou, K. (2021, July 21). Democracy under attack: Challenges of addressing ethical issues of AI and big data for more democratic digital media and societies. *Frontiers in Political Science*, 3.

Christopher, N. (2023, July 5). An Indian politician says scandalous audio clips are AI deepfakes: We had them tested. *Rest of World*. https://restofworld.org/2023/indian-politician-leaked-audio-ai-deepfake/.

Cichy, M. K. (2019, April 16). Does AI have a dirty mind, too? *Medium*. https://medium.com/@marekkcichy/does-ai-have-a-dirty-mind-too-6948430e4b2b.

Cloud, A. (2020, December 7). Media AI: Practice of intelligent video producing. *Medium*. https://medium.datadriveninvestor.com/media-ai-practice-of-intelligent-video-producing-50f658b7d6af.

Coeckelbergh, M. (2020). *AI Ethics*. Cambridge, MA: MIT Press.

Coeckelbergh, M. (2023). How to do robots with words: A performative view of the moral status of humans and nonhumans. *Ethics and Information Technology*, 25(3), 44.

Coeckelbergh, M. & Gunkel, D. J. (2023). ChatGPT: Deconstructing the debate and moving it forward. *AI & Society*, 1–11.

Cohen, J. E. (2019). *Between Truth and Power*. New York: Oxford University Press.

Cook, S. & Palankasova, B. (2023). Communicating research to the public: Creative publics, cross-disciplinary engagement and para-academic practices. In King, B. (ed.), *New Directions for University Museums* (pp. 207–220). A lord cultural resources book. Lanham, MD: Rowman & Littlefield.

Cooper, G. (2020, December 7). Conquering journalism's automation anxiety. *European Journalism Observatory*. https://en.ejo.ch/latest-stories/conquering-journalisms-automation-anxiety.

Couldry, N. (2008). Mediatization or mediation? Alternative understandings of the emergent space of digital storytelling. *New Media & Society*, 10(3), 373–391.

CPA Canada. (2022). Identifying and mitigating bias and mis- and disinformation. *Chartered Professional Accountants Canada*. https/www.ifac.org/system/files/publications/files/Identifying-Mitigating-Bias-Mis-Dis-information-Technology-Ethics-Paper-3.

Czerkawski, B., Brylska, K., & Gackowski, T. (2023). Using AI in scholarly research: Advantages and drawbacks. In Bastiaens, T. (ed.), *Proceedings of EdMedia + Innovate Learning* (pp. 1482–1488). Vienna, Austria: Association for the Advancement of Computing in Education (AACE). Retrieved September 27, 2024 from https://www.learntechlib.org/primary/p/222669/.

Dabis, A. & Csaki, C. (2024). AI and ethics: Investigating the first policy responses of higher education institutions to the challenge of generative AI. *Humanities and Social Sciences Communications*, 11(1), 1–13.

Danaher, J. (2022). Techno-optimism: An analysis, an evaluation and a modest defence. *Philosophy & Technology*, 35, 54.

Danry, V., Leong, J., Pataranutaporn, P., Tandon, P., Liu, Y., Shilkrot, R., Punpongsanon, P., Weissman, T., Maes, P., and Sra, M. (2022, April). AI-generated characters: Putting deepfakes to good use. In Barbosa, S. *et al.* (eds), *CHI Conference on Human Factors in Computing Systems Extended Abstracts* (pp. 1–5). New York: Association for Computing Machinery.

DARPA. (2018). Explainable artificial intelligence (XAI). *DARPA*. https://www.darpa.mil/program/explainable-artificial-intelligence.

Dash, A. & Agres, K. (2024). AI-based affective music generation systems: A review of methods and challenges. *ACM Computing Surveys*, 56(11), 1–34.

David, E. (2024, January 4). OpenAI's news publisher deals reportedly top out at $5 million a year. *The Verge*. https://www.theverge.com/2024/1/4/24025409/openai-training-data-lowball-nyt-ai-copyright.

Davies, H. C., Eynon, R., & Salveson, C. (2021). The mobilisation of AI in education: A Bourdieusean field analysis. *Sociology*, 55(3), 539–560.

Davis, N. (2023, November 13). White faces generated by AI are more convincing than photos, finds survey. *The Guardian*. https://www.theguardian.com/technology/2023/nov/13/white-faces-generated-by-ai-are-more-convincing-than-photos-finds-survey.

de la Tour, A., Soussan, P., Harle, N., Chevalier, R., & Duportet, X. (2017). *From Tech to Deep Tech*. Boston Consulting Group. https://media-publications.bcg.com/from-tech-to-deep-tech.pdf.

de-Lima-Santos, M. F., Yeung, W. N., & Dodds, T. (2024). Guiding the way: A comprehensive examination of AI guidelines in global media. *AI & Society*, 1–19.

de Seta, G. (2021). Huanlian, or changing faces: Deepfakes on Chinese digital media platforms. *Convergence*, 27(4), 935–953.

Demaidi, M. N. (2023). Artificial intelligence national strategy in a developing country. *AI & Society*, 1–13.

Deuze, M. & Beckett, C. (2022). Imagination, algorithms and news: Developing AI literacy for journalism. *Digital journalism*, 10(10), 1913–1918.

Diakopoulos, N. & Johnson, D. (2021). Anticipating and addressing the ethical implications of deepfakes in the context of elections. *New Media & Society*, 23(7), 2072–2098.

Diehl, T. (2023, December 15). MoMA's daydream of progress. *The New York Times*. https://www.nytimes.com/2022/12/15/arts/design/refik-anadol-unsupervised-moma-review.html. (Exhibition data: https://www.moma.org/calendar/exhibitions/5535).

Dwivedi, Y. K., Hughes, L., Ismagilova, E., Aarts, G., Coombs, C., Crick, T., Duan, Y., Dwivedi, R., Edwards, J., Eirug, A., Galanos, V., Vigneswara Ilavarasan, P., Janssen, M., Jones, P., Kar, A.P., Kizgin, H., Kronemann, B., Lal, B., Lucini, B., Medaglia, R., Le Meunier-FitzHugh, K., Le Meunier-FitzHugh, L.C., Misra, S., Mogaji, E., Sharma, S.K., Singh, J.B., Raghavan, V., Raman, R., Rana, N.P., Samothrakis, S., Spencer, J., Tamilmani, K., Tubadji, A., Walton, P., & Williams, M.D. (2021). Artificial Intelligence (AI): Multidisciplinary perspectives on emerging challenges, opportunities, and agenda for research, practice and policy. *International Journal of Information Management*, 57, 101994.

Dwivedi, Y. K., Kshetri, N., Hughes, L., Slade, E.L., Jeyaraj, A., Kumar Kar, A., Baabdullah, A.M., Koohang, A., Raghavan, V., Ahuja, M., Albanna, H., Albashrawi, M.A., Al-Busaidi, A.S., Balakrishnan, J., Barlette, Y., Basu, S., Bose, I., Brooks, L., Buhalis, D., Carter, L., Chowdhury, S., Crick, T., Cunningham, S.W., Davies, G.H., Davison, R.M., Dé, R., Dennehy, D., Duan, Y., Dubey, R., Dwivedi, R., Edwards, J.S., Flavián, C., Gauld, R., Grover, V., Hu, M-C., Janssen, M., Jones, P., Junglas, I., Khorana, S., Kraus, S., Larsen, K.R., Latreille, P., Laumer, S., Malik, F.T., Mardani, A., Mithas, S., Mogaji, E., Horn Nord, J., O'Connor, S., Okumus, F., Pagani, M., Pandey, N., Papagiannidis, S., Pappas, I.O., Pathak, N., Pries-Heje, J., Raman, R., Rana, N.P., Rehm, S-V., Ribeiro-Navarrete, S., Richter, A., Rowe, F., Sarker, S., Carsten Stahl, B., Tiwari, M.K., van der Aalst, W., Venkatesh, V., Viglia, G., Wade, M., Walton, P., Wirtz, J., & Wright, R. (2023). So what if ChatGPT wrote it? Multidisciplinary perspectives on opportunities, challenges and implications of generative conversational AI for research, practice and policy. *International Journal of Information Management*, 71, 102642.

Echauri, G. (2023). Infinite media: The contemporary infinite paradigm in media. *Convergence*, doi:13548565231208135.

Engelberg Center Live. (2024, December 4). Synthetic media: AI and journalism [MP3]. *EC Live*. https://eclive.engelberg.center/episodes/synthetic-media-ai-and-journalism-AGxEocDR.

Epstein, Z., Hertzmann, A. and The Investigators of Human Creativity (2023). Art and the science of generative AI: A deeper dive. *arXiv*. doi:2306.04141.

European Union. (2024/1689). Artificial Intelligence Act [EU AI Act]. https://artificialintelligenceact.eu/.

European Union. (2022/2065). Digital Services Act [DSA]. https://www.eu-digital-services-act.com/.

Exquisite Workers. (2023, May 30). AI surrealism: The world's largest ai art exhibition, NYC, 2023. *Medium*. https://exquisiteworkers.medium.com/ai-surrealism-the-worlds-largest-ai-art-exhibition-2023-8980be9d3e6a.

Fabbri, M. (2023). Social influence for societal interest: A pro-ethical framework for improving human decision making through multi-stakeholder recommender systems. *AI & Society*, 38(2), 995–1002.

Falyuna, N. (2022). Science disinformation as a security threat and the role of science communication in the disinformation society. *Scientia et Securitas*, 1–10.

Falyuna, N., Feher, K., Demeter, M., Szudi, G. & Uscinski, J. E. (2024). Against disinformation: Bridging science and public discourse. *Információs Társadalom*, 24(2), 68–84.

Faraj, S. & Leonardi, P. M. (2022). Strategic organization in the digital age: Rethinking the concept of technology. *Strategic Organization*, 20(4), 771–785.

Feher, K. (2014). What strategies shape digital identity? Exploratory research phase: university students approaching employment [in Hungarian]. *Mediakutato*, Summer, 139–154.

Feher, K. (2016a). Digital mobilization and identity after smart turn. In *Psychology and Mental Health: Concepts, Methodologies, Tools, and Applications* (pp. 1811–1832). IGI Global.

Feher, K. (2016b). *Digitalization and New Media* [in Hungarian]. Budapest: Akademiai Kiado.

Feher, K. (2017). Netframework and the digitalized-mediatized self. *Corvinus Journal of Sociology and Social Policy*, 8(1), 111–126.

Feher, K. (2020). Trends and business models of new-smart-AI (NSAI) media. In *2020 13th CMI Conference on Cybersecurity and Privacy (CMI)-Digital Transformation-Potentials and Challenges (51275)* (pp. 1–6). IEEE.

Feher, K. (2021). Digital identity and the online self: Footprint strategies – An exploratory and comparative research study. *Journal of Information Science*, 47(2), 192–205.

Feher, K. (2022a). *Trust in AI-mediatization: Dynamics of social-cultural diversity and biases*. ECREA 2022: 9th European Communication Conference, Aarhus, October 19–22.

Feher, K. (2022b). *Uncertainties in human-centered information and media technologies*. 6th Conference of the Association of Hungarian American Academicians. May 5–7.

Feher, K. (2022c). Emotion artificial intelligence: Deep mediatised and machine-reflected self-emotions. In Kopecka-Piech, K. & Sobiech, M. (eds), *Mediatisation of Emotional Life* (pp. 41–55). Oxford and New York: Routledge.

Feher, K. (2023a). *Integrative ethics for tech-media industry*. Media Futures Conference, London School of Economics and Political Science, London, June 15–16.

Feher, K. (2023b, April 27). The carbon footprint of AI is already upon us: Competitiveness vs. sustainability? *Carbon Crane*. https://carboncrane.io/blog/post/the-carbon-footprint-of-ai-is-already-upon-us?theme=dark&lang=en.

Feher, K. (2024a). Exploring AI media: Definitions, conceptual model, research agenda. *Journal of Media Business Studies*, 1–24.

Feher, K. (2024b). *Exploring the concept of generative mediatization*. ECREA 2024, 10th European Communication Conference, Ljubljana, 24–27 September.

Feher, K. (2024c). From ChatGPT to AI-driven synthetic worlds: Risks and ethical issues. *Information Society* [in print].

Feher, K. & Katona, A. I. (2021). Fifteen shadows of socio-cultural AI: A systematic review and future perspectives. *Futures*, 132, 102817.

Feher, K. & Katona, A. I. (2023). How do social media machines affect self-concept research? Systematic literature review of the latest trends. *KOME*, 11(1), 2–27.

Feher, K. & Kiraly, O. (2017). Fake news – The dynamics of false news in the media [in Hungarian]. *Szazadveg*, 2.

Feher, K., McKelvey, F., Marinov, R. & Rodgers, L. (2025). *Strategic Foresight in AI Governance: The Archer Model for Responsible AI Futures*. Under Review.

Feher, K. & Veres, Z. (2023). Trends, risks and potential cooperations in the AI development market: Expectations of the Hungarian investors and developers in an international context. *International Journal of Sociology and Social Policy*, 43(1/2), 107–125.

Feher, K., Vicsek, L., & Deuze, M. (2024). Modeling AI trust for 2050: Perspectives from media and info-communication experts. *AI & Society*, 1–14.

Felzmann, H., Villaronga, E. F., Lutz, C., & Tamò-Larrieux, A. (2019). Transparency you can trust: Transparency requirements for artificial intelligence between legal norms and contextual concerns. *Big Data & Society*, 6(1), doi:2053951719860542.

Ferrara, E. (2024). GenAI against humanity: Nefarious applications of generative artificial intelligence and large language models. *Journal of Computational Social Science*, 1–21.

Ferrari, F. & McKelvey, F. (2023). Hyperproduction: A social theory of deep generative models. *Distinktion: Journal of Social Theory*, 24(2), 338–360.

Ferraro, C., Sands, S., Zubcevic-Basic, N., & Campbell, C. (2024). Diversity in the digital age: How consumers respond to diverse virtual influencers. *International Journal of Advertising*, 1–23.

Feuerriegel, S., Hartmann, J., Janiesch, C., & Zschech, P. (2024). Generative AI. *Business & Information Systems Engineering*, 66(1), 111–126.

Fischer, R. & Jarren, O. (2024). The platformization of the public sphere and its challenge to democracy. *Philosophy & Social Criticism*, 50(1), 200–215.

Fisher, E. (2010). Audience labour on social media: Learning from sponsored stories. In Fisher, E. & Fuchs, C. (eds), *Reconsidering Value and Labour in the Digital Age*. New York: Palgrave.

Fjeld, J., Achten, N., Hilligoss, H., Nagy, A., & Srikumar, M. (2020). Principled artificial intelligence: Mapping consensus in ethical and rights-based approaches to principles for AI. *Berkman Klein Center Research Publication*, 2020–1.

Floridi, L. (ed.). (2016). *The Routledge Handbook of Philosophy of Information*. London: Routledge.

Floridi, L. (2023). AI as agency without intelligence: On ChatGPT, large language models, and other generative models. *Philosophy & Technology*, 36(1), 15.

Foley, C. (2022). Why we need to think about diversity and ethics in AI. *IMPACT magazine* (ATSE) [reprint]. https://www.chiefscientist.gov.au/news-and-media/why-we-need-think-about-diversity-and-ethics-ai.

Ford, M. (2018). The truth about AI from the people building it. *Medium*. https://medium.com/@MFordFuture/the-truth-about-ai-from-the-people-building-it-74f95b08b6c8.

Franzke, A. S., Muis, I., & Schafer, M. T. (2021). Data Ethics Decision Aid (DEDA): A dialogical framework for ethical inquiry of AI and data projects in the Netherlands. *Ethics and Information Technology*, 23(3), 551–567.

Fredheim, R. (2023). Virtual manipulation brief 2023/1: Generative AI and its implications for social media analysis. *NATO Strategic Communications Centre of Excellence*. https://stratcomcoe.org/publications/virtual-manipulation-brief-20231-generative-ai-and-its-implications-for-social-media-analysis/286.

Fui-Hoon Nah, F., Zheng, R., Cai, J., Siau, K., & Chen, L. (2023). Generative AI and ChatGPT: Applications, challenges, and AI-human collaboration. *Journal of Information Technology Case and Application Research*, 25(3), 277–304.

Future of Life Institute. (2023, March 22). Pause giant AI experiments: An open letter. https://futureoflife.org/open-letter/pause-giant-ai-experiments/.

Galloway, S. (2022, January 18). The false promise of Web3. *Medium.* https://marker.medium.com/the-false-promise-of-web3-7e6c1a00d4be.

Galloway, S. (2023, June 17). Techno-narcissism. *Medium.* https://medium.com/@profgalloway/techno-narcissism-d2e199c2a002.

Gartner Futures Lab. (2023, April 26). Complexity, chaos and confidence: A tapestry of trends across brave new worlds. *Gartner.* https://www.gartner.com/en/documents/4309299.

Gehl, R. W. & Lawson, S. T. (2022). *Social Engineering: How Crowdmasters, Phreaks, Hackers, and Trolls Created a New Form of Manipulative Communication.* Cambridge, MA: MIT Press.

Gentzkow, M. (2020). *Media and Artificial Intelligence.* https://web.stanford.edu/~gentzkow/research/ai_and_media.pdf.

Gerards, J., Schafer, M. T., Vankan, A. and Muis, I. (2021). Fundamental Rights & Algorithms Impact Assessment (FRAIA). *Government of the Netherlands.* https://www.government.nl/documents/reports/2021/07/31/impact-assessment-fundamental-rights-and-algorithms.

Gonçalves, J., Weber, I., Masullo, G. M., Torres da Silva, M., & Hofhuis, J. (2023). Common sense or censorship: How algorithmic moderators and message type influence perceptions of online content deletion. *New Media & Society*, 25(10), 2595–2617.

Goode, L. (2023, May 26). Where memory ends and generative AI begins. *Wired.* https://www.wired.com/story/where-memory-ends-and-generative-ai-begins/.

Google. (2024). DeepDream. https://deepdreamgenerator.com/.

Google DeepMind. (2023). Gemini: Google's new multimodal AI. https://deepmind.google/technologies/gemini/#capabilities.

De Gregorio, I. (2023, June 16). AI company valued at $260 million: It's only four weeks old. *Medium.* https://medium.com/@ignacio.de.gregorio.noblejas/ai-company-valued-at-260-million-its-only-four-weeks-old-80b7508fcadd.

Grosswiler, P. (2001). Jurgen Habermas: Media ecologist? In *Proceedings of the Media Ecology Association* (Vol. 2, pp. 22–31).

Gu, J. (2024). Responsible generative AI: What to generate and what not. *arXiv.* doi:2404.05783.

Guardian-GPT3. (2020, September 8). A robot wrote this entire article: Are you scared yet, human? *The Guardian.* https://www.theguardian.com/commentisfree/2020/sep/08/robot-wrote-this-article-gpt-3.

Gujral, J. (2023, August 14). The hallucination conundrum in large language models. *LinkedIn.* https://www.linkedin.com/pulse/hallucination-conundrum-large-language-models-rana-gujral/?trackingId=1XstYXbG6fB/ydJNP3Xxig==.

Gunkel, D. J. (2018). *Robot Rights.* Cambridge, MA: MIT Press.

Gupta, A. & Heath, V. (2020, September 14). AI ethics groups are repeating one of society's classic mistakes. *MIT Technology Review.* https://www.technologyreview.com/2020/09/14/1008323/ai-ethics-representation-artificial-intelligence-opinion/.

Guttman, A. (2023, March 11). AI use in media and entertainment – Statistics & facts. *Statista.* https://www.statista.com/topics/10606/artificial-intelligence-use-in-media-and-entertainment/#editorsPicks.

Guzman, A. L. & Lewis, S. C. (2024). What generative AI means for the media industries, and why it matters to study the collective consequences for advertising, journalism, and public relations. *Emerging Media*, doi:27523543241289239.

Hamilton, K. A., Lee, |S. Y., Chung, U. C., Liu, W., & Duff, B. R. (2021). Putting the "me" in endorsement: Understanding and conceptualizing dimensions of self-endorsement using intelligent personal assistants. *New Media & Society*, 23(6), 1506–1526.

Hao, K. (2020, December 4). We read the paper that forced Timnit Gebru out of Google: Here's what it says. *MIT Technology Review*. https://www.technologyreview.com/2020/12/04/1013294/google-ai-ethics-research-paper-forced-out-timnit-gebru/.
Hao, K. (2021, October 20). A deepfake bot is being used to "undress" underage girls. *MIT Technology Review*. https://www.technologyreview.com/2020/10/20/1010789/ai-deepfake-bot-undresses-women-and-underage-girls/.
Hao, K. (2022a, April 19). Artificial intelligence is creating a new colonial world order. *MIT Technology Review*. https://www.technologyreview.com/2022/04/19/1049592/artificial-intelligence-colonialism/.
Hao, K. (2022b, April 2022). A new vision of artificial intelligence for the people. *MIT Technology Review*. https://www.technologyreview.com/2022/04/22/1050394/artificial-intelligence-for-the-people/.
Harvard Carr Center for Human Rights Policy. (2023). Fortifying the truth: Human rights, generative AI, and deepfakes. *YouTube*. https://www.youtube.com/watch?v=DZo4yDeygXI.
Hatterjee, M. & Kern, R. (2023, October 27). Sweeping new Biden order aims to alter the AI landscape. *Politico*. https://www.politico.com/news/2023/10/27/white-house-ai-executive-order-00124067.
Hayles, N. K. (2017). *Unthought: The Power of the Cognitive Nonconscious*. University of Chicago Press.
Heaven, W. D. (2022). Bias isn't the only problem with credit scores – And no, AI can't help. In Martin, K. (ed.), *Ethics of Data and Analytics: Concepts and Cases*. Oxford: CRC Press.
Heaven, W. D. (2023a, September 15). DeepMind's cofounder: Generative AI is just a phase. What's next is interactive AI. *MIT Technology Review*. https://www.technologyreview.com/2023/09/15/1079624/deepmind-inflection-generative-ai-whats-next-mustafa-suleyman/.
Heaven, W. D. (2023b, December 19). These six questions will dictate the future of generative AI. *MIT Technology Review*. https://www.technologyreview.com/2023/12/19/1084505/generative-ai-artificial-intelligence-bias-jobs-copyright-misinformation/.
Heikkila, M. (2023, October). This new data poisoning tool lets artists fight back against generative AI. *MIT Technology Review*. https://www.technologyreview.com/2023/10/23/1082189/data-poisoning-artists-fight-generative-ai/.
Helberger, N., Eskens, S. J., van Drunen, M. Z., Bastian, M. B., & Moller, J. E. (2020). Artificial intelligence – Intelligent politics: Challenges and opportunities for media and democracy. *Council of Europe*.
Helmore, E. (2023, March 17). 'We are a little bit scared': OpenAI CEO warns of risks of artificial intelligence. *The Guardian*. https://www.theguardian.com/technology/2023/mar/17/openai-sam-altman-artificial-intelligence-warning-gpt4.
Hepp, A. (2020). *Deep Mediatization*. Oxford and New York: Routledge.
Hirsch, D. D., Bartley, T., Chandrasekaran, A., Norris, D., Parthasarathy, S., & Turner, P. N. (2020). *Business Data Ethics: Emerging Trends in the Governance of Advanced Analytics and AI*. Ohio State Legal Studies Research Paper.
Hjarvard, S. (2013). *The Mediatization of Culture and Society*. London: Routledge.
Hjarvard, S. (2022). *Media and infrastructure: Addressing the question of power in mediatization research*. Mediatiztaion 3.0 workshop 2022. https://ecrea.eu/event-4756915.
Hobbes, T. (1651). *Leviathan*. Clarendon Press.
Hodgkinson, I. R. & Jackson, T. (2023, August 7). Three steps for businesses to make AI data and compute more sustainable. *OECD.AI Policy Observatory*. https://oecd.ai/en/wonk/three-steps-sustainable-compute-data.
Hoes, E., Aitken, B., Zhang, J., Gackowski, T., & Wojcieszak, M. (2024). Prominent misinformation interventions reduce misperceptions but increase scepticism. *Nature Human Behaviour*, 1–9.

Hoffman, S. G., Joyce, K., Alegria, S., Bell, S. E., Cruz, T. M., Noble, S. U., Shestakofsky, B. & Smith-Doerr, L. (2022). Five big ideas About AI. *Contexts*, 21(3), 8–15.

Howard, P. N. & Kollanyi, B. (2016). Bots, #StrongerIn, and #Brexit: Computational propaganda during the UK–EU referendum. *arXiv*. doi:1606.06356.

Hu, T. (2024). China deploys censors to create socialist AI. *Financial Times*. https://www.ft.com/content/10975044-f194-4513-857b-e17491d2a9e9.

HubSpot & Jasper. (2023). *Using Generative AI to Scale Your Content Operations*. https://offers.hubspot.com/generative-ai-for-content-operations.

IBM. (2023). Tech spend: How will you pay for it? *IBM Institute for Business Value*. https://www.ibm.com/thought-leadership/institute-business-value/en-us/report/ceo-generative-ai/tech-spend.

IEEE. (2024, 30 August). IEEE guide for an architectural framework for explainable artificial intelligence. *IEEE Std 2894-2024*, 1–55.

Iliadis, A., Siapera, E., & Lokot, T. (2023). Decolonising the internet: An introduction to the #AoIR2022 special issue. *Information, Communication & Society*, 26(12), 2369–2375.

Institute for Ethics in AI Oxford. (2023). Generative ethics and the new Bing. *YouTube*. https://www.youtube.com/watch?v=SaSP41BKUkg.

Islam, T., Miron, A., Nandy, M., Choudrie, J., Liu, X., & Li, Y. (2024). Transforming digital marketing with generative AI. *Computers*, 13(7), 168.

Jabłońska, M. R. & Zajdel, R. (2020). Artificial neural networks for predicting social comparison effects among female Instagram users. *PloS one*, 15(2), e0229354.

Jang, W., Kwak, D. H., & Bucy, E. (2022). Knowledge of automated journalism moderates evaluations of algorithmically generated news. *New Media & Society*, doi:14614448221142534.

Jansen, S. C. (2022). *What Was Artificial Intelligence* ? Mediastudies.Press.

Jansson, A. (2015). Using Bourdieu in critical mediatization research. *MedieKultur*, 58: 13–29.

Jee, C. (2019, September 26). 70 countries have now experienced organized disinformation campaigns. *MIT Technology Review*. https://www.technologyreview.com/2019/09/26/102598/70-countries-around-the-world-now-run-organized-disinformation-campaigns/#Echobox=1572026595.

Jenkins, H. (2006). *Convergence Culture: Where Old and New Media Collide*. New York and London: New York University Press.

Jensen, K. B. (2010). *Media Convergence*. New York: Routledge.

Jim The AI Whisperer. (2023, August 16). Is ChatGPT getting dumber? Let's talk about 'AI drift'. *Medium*. https://medium.com/the-generator/is-chatgpt-getting-dumber-lets-talk-about-ai-drift-58f0fc5ebad2.

Jin, S. V. (2023). "To comply or to react, that is the question:" The roles of humanness versus eeriness of AI-powered virtual influencers, loneliness, and threats to human identities in AI-driven digital transformation. *Computers in Human Behavior: Artificial Humans*, 1(2), 100011.

Kantrowitz, A. (2023, May 13). Google, generative search, and the web's uncertain future. *Medium*. https://kantrowitz.medium.com/google-generative-search-and-the-webs-uncertain-future-7d46903d25ba.

Karnouskos, S. (2022). Symbiosis with artificial intelligence via the prism of law, robots, and society. *Artificial Intelligence and Law*, 30(1), 93–115.

Keary, T. (2023, November 8). Is GPT-4 a flop? Taking a closer look at performance challenges and realities. *Technopedia*. https://www.techopedia.com/is-gpt-4-a-flop.

Keenan, S. (2023, October 20). The black GPT: Introducing the AI model trained with diversity and inclusivity in mind. *POCIT*. https://peopleofcolorintech.com/articles/the-black-gpt-introducing-the-ai-model-trained-with-diversity-and-inclusivity-in-mind/?utm_source=

rss&utm_medium=rss&utm_campaign=the-black-gpt-introducing-the-ai-model-trained-with-diversity-and-inclusivity-in-mind.

Kennedy, A. (2023). Generative AI: The evolution of thoughtful online search [a LinkedIn course]. *LinkedIn.* https://www.linkedin.com/learning/generative-ai-the-evolution-of-thoughtful-online-search/how-finding-and-sharing-information-online-has-evolved.

Kenney, M. & Zysman, J. (2016). The rise of the platform economy. *Issues in Science and Technology*, 32(3), 61.

Kerr, A., Barry, M., & Kelleher, J. D. (2020). Expectations of artificial intelligence and the performativity of ethics: Implications for communication governance. *Big Data & Society*, 7(1), doi:2053951720915939.

Klenk, M. (2024). Ethics of generative AI and manipulation: A design-oriented research agenda. *Ethics and Information Technology*, 26(9).

Knott, A., Pedreschi, D., Chatila, R., Chakraborti, T., Leavy, S., Baeza-Yates, R., Eyers, D., Trotman, A., Teal, P., Biecek, P., Russell, S., and Bengio, Y. (2023). Generative AI models should include detection mechanisms as a condition for public release. *Ethics and Information Technology*, 25(4), 55.

Koplin, J. J. (2023). Dual-use implications of AI text generation. *Ethics and Information Technology*, 25(2), 32.

Kotliar, D. M., & Rivnai-Bahir, S. (2023). Contextualizing the ethics of algorithms: A socio-professional approach . *New Media & Society*, 1–21.

Koves, A., Feher, K., Vicsek, L., Fischer, M. (2024). Entangled AI: Artificial intelligence that serves the future. *AI & Society.* doi:10.1007/s00146-024-02037-4.

Kozyrkov, C. (2023a, January 24). What is bias?. *Medium.* https://towardsdatascience.com/what-is-ai-bias-6606a3bcb814.

Kozyrkov, S. (2023b, September 30). AI bias: Good intentions can lead to nasty results. *Medium.* https://kozyrkov.medium.com/ai-bias-good-intentions-can-lead-to-nasty-results-d7b1f309d4b3.

KPMG & University of Queensland. (2023). *Trust in Artificial Intelligence: A Global Study.* https://assets.kpmg.com/content/dam/kpmg/au/pdf/2023/trust-in-ai-global-insights-2023.pdf.

Kramer, M.W. (2018). Uncertainty management. In Farazmand, A. (eds), *Global Encyclopedia of Public Administration, Public Policy, and Governance.* Cham: Springer.

Kruger, M. L. L. (2023, June 6). Risks and ethical considerations of generative AI. *Deloitte.* https://www2.deloitte.com/uk/en/blog/financial-services/2023/risks-and-ethical-considerations-of-generative-ai.html.

Laba, N. (2024). Engine for the imagination? Visual generative media and the issue of representation. *Media, Culture & Society*, doi:01634437241259950.

Laestadius, L., Bishop, A., Gonzalez, M., Illencik, D., & Campos-Castillo, C. (2022). Too human and not human enough: A grounded theory analysis of mental health harms from emotional dependence on the social chatbot Replika. *New Media & Society*, doi:14614448221142007.

Lanier, J. (2014). *Who Owns the Future?*New York: Simon & Schuster.

Lauer, D. (2021). Facebook's ethical failures are not accidental; they are part of the business model. *AI and Ethics*, 1(4), 395–403.

Lee, S. Y., Cho, D., Won, Y. H., & Kim, M. (2022). Special issue on realistic and immersive media technologies. *ETRI Journal*, 44(1), 7–9.

Leo-Liu, J. & Wu-Ouyang, B. (2022). A "soul" emerges when AI, AR, and Anime converge: A case study on users of the new anime-stylized hologram social robot "Hupo". *New Media & Society*, doi:14614448221106030.

Leveson, N. G. (2016). *Engineering a Safer World: Systems Thinking Applied to Safety* (p. 560). Cambridge, MA: The MIT Press.

Lewin, K. (1947). Frontiers in group dynamics: II. Channels of group life; social planning and action research. *Human relations*, 1(2), 143–153.

Lewis, D. (2023, March 21). A psychologist explains how AI and algorithms are changing our lives. *The Wall Street Journal*. https://www.wsj.com/articles/algorithms-ai-humanity-psychology-ebf1364c.

Lewis, D. & Moorkens, J. (2020). A rights-based approach to trustworthy AI in social media. *Social Media+ Society*, 6(3), 2056305120954672.

Lewis, S. C., Sanders, A. K., & Carmody, C. (2019). Libel by algorithm? Automated journalism and the threat of legal liability. *Journalism & Mass Communication Quarterly*, 96(1), 60–81.

Li, F., Larimo, J., & Leonidou, L. C. (2021). Social media marketing strategy: Definition, conceptualization, taxonomy, validation, and future agenda. *Journal of the Academy of Marketing Science*, 49, 51–70.

Li, L., Ma, Z., Fan, L., Lee, S., Yu, H., & Hemphill, L. (2024). ChatGPT in education: A discourse analysis of worries and concerns on social media. *Education and Information Technologies*, 29(9), 10729–10762.

Liao, W., Lu, X., Fei, Y., Gu, Y., & Huang, Y. (2024). Generative AI design for building structures. *Automation in Construction*, 157, 105187.

Lin, C. K. & Jackson, S. J. (2023). From bias to repair: Error as a site of collaboration and negotiation in applied data science work. In *Proceedings of the ACM on Human-Computer Interaction*, 7(CSCW1) (pp. 1–32). New York: Association for Computing Machinery.

Lindgren, H. (2013). Sociotechnical systems as innovation systems in the medical and health domain. In Beuscart-Zéphir, M.C. *et al.* (eds), *Context Sensitive Health Informatics: Human and Sociotechnical Approaches* (pp. 35–40). Amsterdam: IOS Press.

Liu, G. (2022, June 21). The world's smartest artificial intelligence just made its first magazine cover. *Cosmopolitan*. https://www.cosmopolitan.com/lifestyle/a40314356/dall-e-2-artificial-intelligence-cover/.

Lockett, W. (2023, July 29). AI is facing a seriously huge problem. *Medium*. https://medium.com/predict/ai-is-facing-a-seriously-huge-problem-a56733e5ef47.

Luccioni, S., Jernite, Y., & Strubell, E. (2024, June). Power hungry processing: Watts driving the cost of AI deployment?. In *Proceedings of the 2024 ACM Conference on Fairness, Accountability, and Transparency* (pp. 85–99). New York: Association for Computing Machinery.

Lutz, D. (2023, February 1). Uncreative: Will AI eat human creativity alive? *Medium*. https://uxdesign.cc/uncreative-will-ai-eat-human-creativity-alive-a7ef309e1f99.

Makridakis, S. (2017). The forthcoming artificial intelligence (AI) revolution: Its impact on society and firms. *Futures*, 90, 46–60.

Malmborg, F. A., & Trondal, J. (2023). Discursive framing and organizational venues: Mechanisms of artificial intelligence policy adoption. *International Review of Administrative Sciences*, 89(1), 39–58.

Mancera Andrade, J. A., & Teran, L. (2024, June). From GenAI to political profiling avatars: A data-driven approach to crafting virtual experts for voting advice applications. In *Proceedings of the 25th Annual International Conference on Digital Government Research* (pp. 305–311). New York: Association for Computing Machinery.

Manovich, L. (2023, April 20). AI image and generative media: Notes on ongoing revolution. In Manovich, L. & Arielli, E. (eds), *Artificial Aesthetics: Generative AI, Art and Visual Media*. https://manovich.net/content/04-projects/175-artificial-aesthetics/manovich-ai-image-and-generative-media.pdf.

Mansell, R. & Steinmueller, W. E. (2020). *Advanced Introduction to Platform Economics*. Cheltenham: Edward Elgar Publishing.

Mantegna, M. (2024). ARTificial: Why copyright is not the right policy tool to deal with generative AI. *The Yale Law Journal*, 133. https://www.yalelawjournal.org/forum/artificial-why-copyright-is-not-the-right-policy-tool-to-deal-with-generative-ai.

Matsuda, K. (2017). Hyper-reality. *YouTube*. https://www.youtube.com/watch?v=YJg02ivYzSs.

McCarthy, J., Minsky, M. L., Rochester, N., & Shannon, C. E. (2006). A proposal for the Dartmouth summer research project on artificial intelligence, August 31, 1955. *AI magazine*, 27(4), 12.

McStay, A. (2018). *Emotional AI: The Rise of Empathic Media*. London: Sage.

McLuhan, M. (1964). *Understanding Media: The Extensions of Man*. New York: McGraw-Hill.

Megahed, F. M., Chen, Y. J., Ferris, J. A., Knoth, S., & Jones-Farmer, L. A. (2024). How generative AI models such as ChatGPT can be (mis) used in SPC practice, education, and research? An exploratory study. *Quality Engineering*, 36 (2), 287–315.

Mignano, M. (2022, July 27). The end of social media. *Medium*. https://mignano.medium.com/the-end-of-social-media-a88ffed21f86.

Miller, G. (2023, December 8). US senate AI 'Insight Forum' tracker. *Tech Policy Press*. https://www.techpolicy.press/us-senate-ai-insight-forum-tracker/.

Milliere, R. (2022, May 11). Deep learning and synthetic media. *Arxiv*. https://arxiv.org/abs/2205.05764.

Mir, A. (2023, July 6). The final merging of humans and their media has begun. *Discourse*. https://www.discoursemagazine.com/ideas/2023/07/06/the-final-merging-of-humans-and-their-media-has-begun/.

Mirksy, Y., Brodt, O., Cohen, J., Levy, R., & Blokh, I. (2019). Live deepfake. *Journal of Creating Gans and Artificial Intelligence*. https://hakita.co.il/upload/posts/20e474ca078aefd4d1aede450994a944.pdf.

MIT IDE. (2023a, August 28). 4 Questions to ask before swapping human labor for AI. *Medium*. https://medium.com/mit-initiative-on-the-digital-economy/4-questions-to-ask-before-swapping-human-labor-for-ai-29e96ada73db.

MIT IDE. (2023b, August 18). Three ways that AI progress could stall. *Medium*. https://medium.com/mit-initiative-on-the-digital-economy/three-reasons-that-ai-progress-could-stall-e0c9af463a9e.

MIT Technology Review Insights. (2023, July 18). The great acceleration: CIO perspectives on generative AI. *MIT Technology Review*. https://www.technologyreview.com/2023/07/18/1076423/the-great-acceleration-cio-perspectives-on-generative-ai/.

Mittelstadt, B. (2019). Principles alone cannot guarantee ethical AI. *Nature Machine Intelligence*, 1(11), 501–507.

Mollick, E. (2024). *Co-Intelligence*. London: Random House.

Monge, J. C. (2023, November 17). This AI-powered presentation maker is mind-blowing. *Medium*. https://generativeai.pub/this-ai-powered-presentation-maker-is-mind-blowing-cdf927a058bf.

Moore, A. D. (2017). Privacy, neuroscience, and neuro-surveillance. *Res Publica*, 23, 159–177.

Morris, M. R. & Brubaker, J. R. (2024). Generative ghosts: Anticipating benefits and risks of AI afterlives. *arXiv*. doi:2402.01662.

Muhlhoff, R. (2020). Human-aided artificial intelligence: Or, how to run large computations in human brains? Toward a media sociology of machine learning. *New Media & Society*, 22(10), 1868–1884.

Musi, E., Aguilar, E. E. G., & Federico, L. (2024). Botlitica: A generative AI-based tool to assist journalists in navigating political propaganda campaigns. *Studies in Communication Sciences*, 24(1), 161–169.

Narayanan, A. & Kapoor, S. (2023, June 26). Generative AI companies must publish transparency reports. *Knight First Amendment Institute, Columbia University*. https://knightcolumbia.org/blog/generative-ai-companies-must-publish-transparency-reports.

Narayan, J. (2022, June 15). *National AI strategies – Implementation and the road ahead*. World AI Summit. London, UK.

Narayanan, S. (2018). Tutorial: 21 fairness definitions and their politics. *YouTube*. https://www.youtube.com/watch?v=jIXIuYdnyyk.

Natale, S. (2022). *What is being automated, anyway? Interrogating automated sociality through histories*. ECREA 2022 9th European Communication Conference. AARHUS, Denmark, 19–22 October.

Natale, S. & Cooke, H. (2021). Browsing with Alexa: Interrogating the impact of voice assistants as web interfaces. *Media, Culture & Society*, 43(6), 1000–1016.

Nature. (2023, June 7). Why Nature will not allow the use of generative AI in images and videos. *Nature*. https://www.nature.com/articles/d41586-023-01546-4.

Nelson, A. (2023). Ethics in AI annual lecture with Alondra Nelson: University of Oxford, Institute for Ethics in AI. https://www.oxford-aiethics.ox.ac.uk/event/ethics-ai-annual-lecture-alondra-nelson-person-only-event-oxford-uk.

Neuhauser, L., Stamm, F. I., Lemmerich, F., Schaub, M. T., & Strohmaier, M. (2021). Simulating systematic bias in attributed social networks and its effect on rankings of minority nodes. *Applied Network Science*, 6(1), 86.

Newel, B. (2021, May 28). Daniel Kahneman on 'noise' – The flaw in human judgement harder to detect than cognitive bias. *The Conversation*. https://theconversation.com/daniel-kahneman-on-noise-the-flaw-in-human-judgement-harder-to-detect-than-cognitive-bias-160525.

Nguyen, T., Jump, A., & Casey, D. (2023). Emerging tech impact radar. *Gartner Research*. https://www.gartner.com/en/doc/emerging-technologies-and-trends-impact-radar-excerpt.

Nielsen, M. A. & Chuang, I. L. (2010). *Quantum Computation and Quantum Information: 10th Anniversary Edition*. Cambridge University Press.

Nishant, R., Kennedy, M., & Corbett, J. (2020). Artificial intelligence for sustainability: Challenges, opportunities, and a research agenda. *International Journal of Information Management*, 53, 102104.

Noain-Sanchez, A. (2022). Addressing the impact of artificial intelligence on journalism: The perception of experts, journalists and academics. *Communication and Society*, 35 (3), 105–121.

Noorman, M. & Swierstra, T. (2023). Democratizing AI from a sociotechnical perspective. *Minds and Machines*, 33, 563–586.

Nouri, S. (2023, December 22). Google dropped the ChatGPT killer!? *Medium*. https://medium.com/@gdscadgitm/did-google-just-release-chatgpt-killer-6dd3a6e3f462.

Noyes, K. (2024, October 10). *FBI Approach to emerging and disruptive technologies: AI*. World Summit AI. Amsterdam, Netherlands.

OECD. (2020). *Good Practice Principles for Data Ethics in the Public Sector*. https://www.oecd-ilibrary.org/oecd-good-practice-principles-for-data-ethics-in-the-public-sector_caa35b76-en.pdf?itemId=%2Fcontent%2Fpaper%2Fcaa35b76-en&mimeType=pdf.

OECD. (2023). *AI Language Models: Technological, Socio-Economic and Policy Considerations*. OECD Digital Economy Papers, 352. Paris: OECD Publishing.

OECD AI. (2024). National strategies, agendas and plans. *OECD AI Policy Observatory*. https://oecd.ai/en/dashboards/policy-instruments/National_strategies_agendas_and_plans.

Opal, E. (2021). Reducing harm on social media: Research & design ideas. https://www.youtube.com/watch?v=F4pAVXbCQoU.

Orf, D. (2023, October 20). A new study says AI is eating its own tail. *Popular Mechanics*. https://www.popularmechanics.com/technology/a44675279/ai-content-model-collapse/#.

Orr, W. & Crawford, K. (2024). The social construction of datasets: On the practices, processes, and challenges of dataset creation for machine learning. *New Media & Society*, 26(9), 4955–4972.

Pacey, A. (1985). *The Culture of Technology*. Cambridge, MA: MIT press.

Packin, N. G. (2021). Disability discrimination using artificial intelligence systems and social scoring: Can we disable digital bias? *Journal of International and Comparative Law*, 8, 487.

Panetta, F., Burgund, H., Ben-Ami, O., Serdiuk, A., Reaber, G., & Bielievtsov, D. (2019). *In event of moon disaster*. SIGGRAPH Asia 2020 Computer Animation Festival.

Papp, L. (2023). *Digital Immune System*. BusinessFest, Budapest, June 8.

Paris, B. S., Cath, C., & West, S. M. (2023). Radical infrastructure: Building beyond the failures of past imaginaries for networked communication. *New Media & Society*, doi:14614448231152546.

Pariser, E. (2011). *The Filter Bubble: What the Internet is Hiding from You*. New York: Penguin.

Pavel, V. (2022, 17 November). Rethinking data and rebalancing digital power. *Ada Lovelace Institute*. https://www.adalovelaceinstitute.org/report/rethinking-data/.

Perrigo, B. (2023, January 18). Exclusive: OpenAI used Kenyan workers on less than $2 per hour to make ChatGPT less toxic. *Time*. https://time.com/6247678/openai-chatgpt-kenya-workers/.

Pew Research Center and Elon University. (2022). The future of the metaverse. *Elon University*. https://www.elon.edu/u/imagining/surveys/xiv-2022/future-of-metaverse-web3-2040/.

Pierson, J., Kerr, A., Robinson, C., Fanni, R., Steinkogler, V., Milan, S., & Zampedri, G. (2023). Governing artificial intelligence in the media and communications sector. *Internet Policy Review*, 12(1).

Posetti, J. & Matthews, A. (2018). A short guide to the history of 'fake news' and disinformation. *International Center for Journalists*. https://www.icfj.org/sites/default/files/2018-07/A.

Potter, W. (2024,July 23). An academic publisher has struck an AI data deal with Microsoft – without their authors' knowledge. *The Conversation*. https://theconversation.com/an-academic-publisher-has-struck-an-ai-data-deal-with-microsoft-without-their-authors-knowledge-235203.

Pragmatic Coders. (2023, December 10). AI predictions: Top 13 AI trends for 2024. *Medium*. https://medium.com/@pragmaticcoders/ai-predictions-top-13-ai-trends-for-2024-4158d23efd78.

Prasad, R. (2020, December 28). The Turing Test is obsolete: It's time to build a new barometer for AI. *Fast Company*. https://www.fastcompany.com/90590042/turing-test-obsolete-ai-benchmark-amazon-alexa.

Purdy, M. (2019, November 18). The risks of using AI to interpret human emotions. *Harvard Business Review*. https://hbr.org/2019/11/the-risks-of-using-ai-to-interpret-human-emotions.

Qadri, R., Shelby, R., Bennett, C. L., & Denton, E. (2023, June). AI's regimes of representation: A community-centered study of text-to-image models in South Asia.

Proceedings of the 2023 ACM Conference on Fairness, Accountability, and Transparency (pp. 506–517). New York: Association for Computing Machinery.

Qi, Z. (2020, November 13). The future of AI: Is there a place for ambiguity? *Medium.* https://medium.com/swlh/the-future-of-ai-is-there-a-place-for-ambiguity-33df0b65bc59.

Qiu, J. L. (2023). The return of billiard balls? US–China tech war and China's state-directed digital capitalism. *Javnost – The Public*, 30(2), 197–217.

Rabowsky, B. (2023). Applications of generative AI to media. *SMPTE Motion Imaging Journal*, 132(8), 53–57.

Rahwan, I. (2018). Society-in-the-loop: Programming the algorithmic social contract. *Ethics and Information Technology*, 20(1), 5–14.

Rahwan, I., Cebrian, M., Obradovich, N., Bongard, J., Bonnefon, J. F., Breazeal, C., Crandall, J.W., Christakis, N.A., Couzin, I.D., Jackson, M.O., Jennings, N.R., Kamar, E., Kloumann, I., Larochelle, H., Lazer, D., McElreath, R., Mislove, A., Parkes, D.C., Pentland, A., Roberts, M.E., Shriff, A., Tenenbaum, J., and Wellman, M. (2019). Machine behaviour. *Nature*, 568, 477–486.

Raley, R. & Rhee, J. (2023). Critical AI: A field in formation. *American Literature*, 95 (2), 185–204.

Ramani, A. & Wang, Z. (2023). Why transformative artificial intelligence is really, really hard to achieve. *The Gradient*. https://thegradient.pub/why-transformative-artificial-intelligence-is-really-really-hard-to-achieve/.

Randieri, C. (2025, January 3). Agentic AI: A new paradigm in autonomous artificial intelligence. *Forbes*. https://www.forbes.com/councils/forbestechcouncil/2025/01/03/agentic-ai-a-new-paradigm-in-autonomous-artificial-intelligence/.

Reisach, U. (2021). The responsibility of social media in times of societal and political manipulation. *European Journal of Operational Research*, 291(3), 906–917.

Reuters. (2021, December 13). AI synthetic media tech enters perilous phase with deepfakes. *Deccan Herald*. https://www.deccanherald.com/technology/ai-synthetic-media-tech-enters-perilous-phase-with-deepfakes-1060293.html.

Riedl, M. O. (2019). Human-centered artificial intelligence and the future of work. *Journal of Artificial Intelligence Research*, 63, 723–731.

Roberts, S. T. (2016). Commercial content moderation: Digital laborers' dirty work. *Western Libraries*. https://ir.lib.uwo.ca/commpub/12/.

Rodgers, W. & Nguyen, T. (2022). Advertising benefits from ethical artificial intelligence algorithmic purchase decision pathways. *Journal of Business Ethics*, 178(4), 1043–1061.

Rosenberg, L. (2022, December 27). Mind control in the metaverse. *Medium*. https://medium.com/predict/mind-control-in-the-metaverse-48dfbd88c2ae.

Rourke, M. (2021, February 27). Emerging new business models for news media innovation. *Innovation*. https://innovation.media/insights/emerging-new-business-models-for-news-media.

Rouse, M. (2023). Generative AI. *Technopedia*. https://www.techopedia.com/definition/34633/generative-ai.

Rousseau, J. J. (1762). *The Social Contract*. London: Penguin Classics.

Russell, S. & Norvig, P. (2020). *Artificial Intelligence: A Modern Approach*. London: Pearson.

Sætra, H. S. (2023). Generative AI: Here to stay, but for good? *Technology in Society*, 75, 102372.

Sanjeev, A. (2023a, March 17). AI's black swans: Unforeseen consequences looming. *Medium*. https://uxdesign.cc/ais-black-swans-unforeseen-consequences-looming-a4b28ef79004.

Sanjeev, A. (2023b, March 17). Artificially intelligent sociopaths: The dark reality of malicious AI. *Medium*. https://medium.com/user-experience-design-1/ai-sociopaths-and-their-unintended-consequences-bf6f19012e4f.

Sanjeev, A. (2023c, March 17). Indistinguishable AI content is crippling our perception. *Medium*. https://uxdesign.cc/indistinguishable-ai-content-is-crippling-our-perception-f2b24c7cd311.

Saputra, M. G. P. (2017). Game history of Raden Patah using Fuzzy Sugeno algorithm. *IEESE International Journal of Science and Technology*, 6(1), 15.

Sartori, L. & Theodorou, A. (2022). A sociotechnical perspective for the future of AI: Narratives, inequalities, and human control . *Ethics and Information Technology*, 24(1), 4.

Schaake, M. (2024). *The Tech Coup: How to Save Democracy from Silicon Valley*. Princeton University Press.

Schaffer, M. T. (2022). *Not merely studying but also building the digital society! How socially engaged research makes a difference*. A keynote lecture. ECREA 2022 9th European Communication Conference, Aarhus, Denmark, 19–22 October.

Schaffer, M. T. (2024). Inside datafication: Entrepreneurial research for investigating emerging data practices. In Schaer, M.T. *et al.* (eds), *Collaborative Research in the Datafied Society: Methods and Practices of Investigation and Intervention* (pp. 61–86). Amsterdam University Press.

Schneider, J. (2024). Explainable generative AI (GenXAI): A survey, conceptualization, and research agenda. *Artificial Intelligence Review*, 57(11), 289.

Schwencke, K. (2014, March 18). Robot writes LA Times earthquake breaking news article. *BBC News*. https://www.bbc.com/news/technology-26614051.

Selvarajan, S., Manoharan, H., Khadidos, A. O., Khadidos, A. O., Shankar, A., Maple, C., & Singh, S. (2024). Generative artificial intelligence and adversarial network for fraud detections in current evolutional systems. *Expert Systems*, e13740.

Sezgin, E. & McKay, I. (2024). Behavioral health and generative AI: A perspective on future of therapies and patient care. *npj Mental Health Research*, 3(1), 25.

Shumailov, I., Shumaylov, Z., Zhao, Y., Papernot, N., Anderson, R., & Gal, Y. (2024). AI models collapse when trained on recursively generated data. *Nature*, 631, 755–759.

Siffels, L., van den Berg, D., Schafer, M. T., Muis, I. (2022). Public values and technological change: Mapping how municipalities grapple with data ethics. In Hepp, A. *et al.* (eds), *New Perspectives in Critical Data Studies: The Ambivalences of Data Power*. Cham: Palgrave.

Sigfrids, A., Leikas, J., Salo-Pontinen, H., & Koskimies, E. (2023). Human-centricity in AI governance: A systemic approach. *Frontiers in Artificial Intelligence*, 6, 976887.

Shah, A. (2024). Media and artificial intelligence: Current perceptions and future outlook. *Academy of Marketing Studies Journal*, 28(2).

Shaikh, S. J. & Moran, R. E. (2022). Recognize the bias? News media partisanship shapes the coverage of facial recognition technology in the United States. *New Media & Society*, doi:14614448221090916.

Shane, T., Saltz, E., & Leibowicz, C. (2021, May 12). From deepfakes to TikTok filters: How do you label AI content? *Nieman Lab*. https://www.niemanlab.org/2021/05/from-deepfakes-to-tiktok-filters-how-do-you-label-ai-content/.

Shao, C., Ciampaglia, G. L., Varol, O., Yang, K. C., Flammini, A., & Menczer, F. (2018). The spread of low-credibility content by social bots. *Nature Communications*, 9(1), 1–9.

Sibert, L. C. (2020). Uncertainty for meaningful human control. *TU Delft*. https://www.tudelft.nl/aitech/projects/uncertainty-for-meaningful-human-control.

Signer, B., Roels, R., van Barlingen, R., & Willems, B. (2021, August). Back to the future: Bringing original hypermedia and cross-media concepts to modern desktop environments. In *Proceedings of the 32nd ACM Conference on Hypertext and Social Media* (pp. 277–282). New York: Association for Computing Machinery.

Siles, I., Valerio-Alfaro, L., & Melendez-Moran, A. (2022). Learning to like TikTok … and not: Algorithm awareness as process. *New Media & Society*, doi:14614448221138973.

Sinatra, G. & Hofer, B.K. (2023, May 24). ChatGPT and other generative AI could foster science denial and misunderstanding – Here's how you can be on alert. *The Conversation*. https://theconversation.com/chatgpt-and-other-generative-ai-could-foster-science-denial-and-misunderstanding-heres-how-you-can-be-on-alert-204897.

Sison, A. J. G., Daza, M. T., Gozalo-Brizuela, R., & Garrido-Merchan, E. C. (2023). ChatGPT: More than a weapon of mass deception, ethical challenges and responses from the human-centered artificial intelligence (HCAI) perspective. *arXiv*. doi:2304.11215.

Sloane, M. (2019, June 17). Co-opting AI. *Public Books*. https://www.publicbooks.org/co-opting-ai/.

Sloane, M., Solano-Kamaiko, I. R., Yuan, J., Dasgupta, A., & Stoyanovich, J. (2023). Introducing contextual transparency for automated decision systems. *Nature Machine Intelligence*, 5(3), 187–195.

Smith, T. (2023, December 11). Why humor is the perfect benchmark for generative AI. *Medium*. https://medium.com/the-generator/can-generative-ai-be-funny-27a377feb6e9.

Snapp, M. (2021, October 7). New steps to preserve and protect journalism and local newsrooms. *Microsoft on the Issues*. https://blogs.microsoft.com/on-the-issues/2020/10/07/local-news-pilot-accountguard-defending-democracy/.

Sourbati, M. & Behrendt, F. (2021). Smart mobility, age and data justice. *New Media & Society*, 23(6), 1398–1414.

Spichak, S. (2021, April 21). Frog cells self-assemble into novel organisms called Xenobots. *Medium*. https://medium.com/age-of-awareness/frog-cells-self-assemble-into-novel-organisms-called-xenobots-401676647712.

Springs. (2024,September 19). How many generative AI startups are there: Latest statistics in 2024. *Springs*. https://springsapps.com/knowledge/how-many-generative-ai-startups-are-there-latest-statistics-in-2024#:~:text=The%20global%20market%20size%20for%20generative%20AI%20solutions%20in%202023.

Stahl, B. C. & Eke, D. (2024). The ethics of ChatGPT – Exploring the ethical issues of an emerging technology. *International Journal of Information Management*, 74, 102700.

Stahl, B. C., Schroeder, D., & Rodrigues, R. (2023). *Ethics of Artificial Intelligence: Case Studies and Options for Addressing Ethical Challenges*. Cham: Springer Nature.

Stanford AI Index. (2023). *Ethical Challenges of AI Applications*. https://aiindex.stanford.edu/wp-content/uploads/2021/03/2021-AI-Index-Report-_Chapter-5.pdf.

Stavola, J. & Choi, K. S. (2023). Victimization by deepfake in the metaverse: Building a practical management framework. *International Journal of Cybersecurity Intelligence & Cybercrime*, 6(2), 2.

Steinhoff, J. (2022). Toward a political economy of synthetic data: A data-intensive capitalism that is not a surveillance capitalism? *New Media & Society*, 26(6), 3290–3306.

Sterne, J. & Razlogova, E. (2019). Machine learning in context, or learning from LANDR: artificial intelligence and the platformization of music mastering. *Social Media+ Society*, 5(2), doi:2056305119847525.

Strickland, L. & Lewis, H. R. (2022). *Leibniz on Binary: The Invention of Computer Arithmetic*. Cambridge, MA: MIT Press.

Sundar, S. S. (2020). Rise of machine agency: A framework for studying the psychology of human–AI interaction (HAII). *Journal of Computer-Mediated Communication*, 25 (1), 74–88.

Taleb, N. (2005). *The Black Swan: Why Don't We Learn That We Don't Learn.* New York: Random House.

TED Salon. (2020). 6 big ethical questions about the future of AI. *TED.* https://www.ted.com/talks/genevieve_bell_6_big_ethical_questions_about_the_future_of_ai/transcript.

The Ford Foundation. (2023, August 25). Ford Foundation Gallery presents What Models Make Worlds: Critical Imaginaries of AI. https://www.fordfoundation.org/news-and-stories/news-and-press/news/ford-foundation-gallery-presents-what-models-make-worlds-critical-imaginaries-of-ai/.

The New York Times R&D. (2021). Spatial journalism. *The New York Times.* https://rd.nytimes.com/research/spatial-journalism/.

The White House. (2023/14110). Executive order on the safe, secure, and trustworthy development and use of artificial intelligence. https://www.whitehouse.gov/briefing-room/presidential-actions/2023/10/30/executive-order-on-the-safe-secure-and-trustworthy-development-and-use-of-artificial-intelligence/.

Thimm, C., Engelhardt, P., Nehls, P., Hens, J., Biehler, M., Hermanns, C., Huzel, P., Bennewitz, M., Mücke, J., & Dengler, N. (2024, June). Trust, (dis) comfort, and voice quality: Anthropomorphism in verbal interactions with NAO. In *Proceedings of the International Conference on Human-Computer Interaction* (pp. 405–412). Cham: Springer Nature.

Thomson, T. J., Thomas, R. J., & Matich, P. (2024). Generative visual AI in news organizations: Challenges, opportunities, perceptions, and policies. *Digital Journalism*, 1–22.

Tollerer, T. (2023,October 26). Leonardo.ai is the new game-changer in the AI world—It puts MidJourney to shame! *Medium.* https://medium.com/@ThomasTTollerer/leonardo-ai–is-the-new-game-changer-in-the-ai-world-it-puts-midjourney-to-shame-8b9a3a55960d.

Towards AI Editorial Team. (2021, August 8). Generative AI and GANs. *Towards AI.* https://towardsai.net/p/l/generative-ai-gans.

Trattner, C., Jannach, D., Motta, E., Costera Meijer, I., Diakopoulos, N., Elahi, M., Opdahl, B.T., Borch, N., Fjeld, M., Øvrelid, L., De Smedt, K., & Moe, H. (2022). Responsible media technology and AI: Challenges and research directions. *AI and Ethics*, 2(4), 585–594.

Tsalakanidou, F., Papadopoulos, S., Mezaris, V., Kompatsiaris, I., Gray, B., Tsabouraki, D., Kalogerini, M., Negro, F., Montagnuolo, M., de Vos, J., van Kemenade, P., Gravina, D., Mignot, R., Ozerov, A., Schnitzler, F., Garcia-Saez, A., Yannakakis, G. N., Liapis, A., & Kostadinov, G. (2021). The AI4Media project: Use of next-generation artificial intelligence technologies for media sector applications. In *Proceedings of Artificial Intelligence Applications and Innovations: 17th IFIP WG 12.5 International Conference, AIAI 2021, Hersonissos, Crete, Greece, June 25–27, 2021* (pp. 81–93). Cham: Springer International Publishing.

Tsamados, A. (2021). Artificial intelligence for social good: Avoiding the solutionist trap. *Eliamep.* https://www.eliamep.gr/en/publication/τεχνητή-νοημοσύνη-για-το-κοινωνικό-όφ/.

Tzezana, R. (2017, February 28). The three AI waves that will shape the future. *Curating the Future.* https://curatingthefuture.com/2017/02/28/ai_future_darpa/.

UAL. (2024, September 18). Charismatic project: AI-driven storytelling for TV & film industries. *University of the Arts London.* https://www.arts.ac.uk/about-ual/press-office/stories/Charismatic-project-AI-driven-storytelling-for-TV-film-industries.

Udupa, S., Maronikolakis, A., and Wisiorek, A. (2023). Ethical scaling for content moderation: Extreme speech and the (in)significance of artificial intelligence. *Big Data & Society*, 10(1).

USC. (2024). Media as sociotechnical systems. *University of Southern California*. https://www.uscmasts.org/.

Utopia. (2023). Ethical AI: AI can be a powerful tool. https://www.utopiaanalytics.com/ethical-ai.

van Es, K. & Nguyen, D. (2024). "Your friendly AI assistant": The anthropomorphic self-representations of ChatGPT and its implications for imagining AI. *AI & Society*.

Velkova, J. & Plantin, J. C. (2023). Data centers and the infrastructural temporalities of digital media: An introduction. *New Media & Society*, 25(2), 273–286.

Verdegem, P. (ed.) (2021). *AI for Everyone?: Critical Perspectives* (p. 310). University of Westminster Press. https://westminsterresearch.westminster.ac.uk/download/ca5c488ee2f3ecadb64aa308c5b69312736ea2f17bb8a9907881b14306bed138/2920707/ai-for-everyone.pdf.

Vermeulen, T. & Van den Akker, R. (2010). Notes on metamodernism. *Journal of Aesthetics & Culture*, 2(1) [article no. 5677].

Veszelszki, A. (2017). Linguistic and non-linguistic elements in detecting (Hungarian) fake news. *Acta Universitatis Sapientiae, Communicatio* (4), 7–35.

Vicsek, L., Nguyen, D. & Feher, K. (2025). Disappointment or hype? Framing costs and benefits of artificial intelligence in the US elite news media (under review).

Vicsek, L., Pinter, R., & Bauer, Z. (2024). Shifting job expectations in the era of generative AI hype – Perspectives of journalists and copywriters. *International Journal of Sociology and Social Policy*. 9 July.

Vilaca, M. M., Karasinski, M., & Candiotto, K. B. B. (2024). Reflections on the future of artificial intelligence: An interview with Luciano Floridi. *Filosofia Unisinos*, 25(1), e25115.

Vincent, J. (2023, February 17). Introducing the AI mirror test, which very smart people keep failing. *The Verge*. https://www.theverge.com/23604075/ai-chatbots-bing-chatgpt-intelligent-sentient-mirror-test.

Vincent, J. (2021, December 6). This AI art app is a glimpse at the future of synthetic media. *The Verge*. https://www.theverge.com/2021/12/6/22820106/ai-art-app-dream-synthetic-media-wombo.

Vinchon, F., Lubart, T., Bartolotta, S., Gironnay, V., Botella, M., Bourgeois-Bougrine, S., Burkhardt, J-M., Bonnardel, N., Corazza, G., Glăveanu, V., Hanchett Hanson, M., Ivcevic, Z., Karwowski, M., Kaufman, J.C., Okada, T., Reiter-Palmon, R. & Gaggioli, A. (2023). Artificial intelligence & creativity: A manifesto for collaboration. *The Journal of Creative Behavior*, 57(4).

Wang, S., Ning, Z., Truong, A., Dontcheva, M., Li, D., & Chilton, L. B. (2024, July). PodReels: Human-AI co-creation of video podcast teasers. In *Proceedings of the 2024 ACM Designing Interactive Systems Conference* (pp. 958–974). New York: Association for Computing Machinery.

Wang, W. (2023, June 22). Calculating political bias and fighting partisanship with AI. *The Bipartisan Press*. https://www.thebipartisanpress.com/politics/calculating-political-bias-and-fighting-partisanship-with-ai/.

Wang, Y. (2021, May). The application of artificial intelligence in Chinese news media. In *Proceeding of 2021 2nd International Conference on Artificial Intelligence and Information Systems* (pp. 1–4). New York: Association for Computing Machinery.

Webb, A. (2021, December 1). Welcome to the (synthetic) meatspace. *Wired*. https://www.wired.com/story/synthetic-meatspace-food-health/.

Weikmann, T. & Lecheler, S. (2023). Visual disinformation in a digital age: A literature synthesis and research agenda. *New Media & Society*, 25(12), 3696–3713.

Winiarska-Brodowska, M. & Feher, K. (2025). Socio-Cultural Artificial Intelligence (SCAI) and journalism: A transdisciplinary perspective in media and communication studies. In

Brzezinski, D. *et al.* (eds), *Algorithms, Artificial Intelligence and Beyond: Theorizing Society and Culture of the 21st Century*. Oxford and New York: Routledge.

Wirtschafter, V. & Dews, F. (2023, October 25). Parsing disinformation in the Israel-Hamas conflict. *Brookings*. https://www.brookings.edu/articles/parsing-disinformation-in-the-israel-hamas-conflict/.

Wischnewski, M., Ngo, T., Bernemann, R., Jansen, M., & Kramer, N. (2022). "I agree with you, bot!" How users (dis)engage with social bots on Twitter. *New Media & Society*, doi:14614448211072307.

Wong, Y. N., Jones, R., Das, R., & Jackson, P. (2023). Conditional trust: Citizens' council on data-driven media personalisation and public expectations of transparency and accountability. *Big Data & Society*, 10(2), doi:20539517231184892.

Wooldridge, M. & Jennings, N. R. (1995). Intelligent agents: Theory and practice. *The Knowledge Engineering Review*, 10(2), 115–152.

Woolley, S. C. & Howard, P. N. (2016). Political communication, computational propaganda, and autonomous agents: Introduction. *International Journal of Communication*, 10.

World Economic Forum. (2023, June 14). *The Presidio Recommendations on Responsible Generative AI*. https://www.weforum.org/publications/the-presidio-recommendations-on-responsible-generative-ai/.

Wu, T. (2018). Blind spot: The attention economy and the law. *Antitrust*.

Wu, Y., Kim, K. J., & Mou, Y. (2022). Minority social influence and moral decision-making in human–AI interaction: The effects of identity and specialization cues. *New Media & Society*, doi:14614448221138072.

Wulf, V., Pipek, V., Randall, D., Rohde, M., Schmidt, K., & Stevens, G. (eds). (2018). *Socio-Informatics*. Oxford University Press.

Yadlin-Segal, A. & Oppenheim, Y. (2021). Whose dystopia is it anyway? Deepfakes and social media regulation convergence. *The International Journal of Research into New Media Technologies*, 27(1), 38.

Yu, W., Payton, B., Sun, M., Jia, W., & Huang, G. (2023). Toward an integrated framework for misinformation and correction sharing: A systematic review across domains. *New Media & Society*, 25(8), 2241–2267.

Zakrzewski, C. & Faiola, A. (2023, November 1). Global adversaries and allies reach first agreement on containing AI risks. *The Media Institute*. https://www.mediainstitute.org/2023/11/01/global-adversaries-allies-reach-first-agreement-on-containing-ai-risks/.

Zakrzewski, C., Tiku, N., Lima-Strong, C., Oremus, W., Colvin, R., Hohmann, J., Verma, P. and Lerman, R. (2023, May 16). OpenAI CEO tells senate that he fears AI's potential to manipulate views. *Washington Post*. https://www.washingtonpost.com/technology/2023/05/16/ai-congressional-hearing-chatgpt-sam-altman/.

Zeineddine, K. (2020, September 24). The interplay between artificial intelligence and uncertainty. *Medium*. https://medium.com/swlh/the-interplay-between-artificial-intelligence-and-uncertainty-a19e01197230.

Zelenkauskaite, A. (2022). *Creating Chaos Online: Disinformation and Subverted Post-Publics* (p. 318). University of Michigan Press.

Zelizer, B. (2021). Why journalism's default neglect of temporality is a problem. *Media, Culture & Society*, 43(7), 1213–1229.

Zhang, Y. (2022). Algorithmic photography: A case study of the Huawei Moon Mode controversy. *Media, Culture & Society*, 44(4), 690–705.

Zhang, W., Lai, K. H., & Gong, Q. (2024). The future of the labor force: Higher cognition and more skills. *Humanities and Social Sciences Communications*, 11(1), 1–9.

Zuboff, S. (2019). *The Age of Surveillance Capitalism: The Fight for a Human Future at the New Frontier of Power*. London: Profile Books.

INDEX

Page references for terms defined in the Glossary are indicated in bold.

MIX
Papier | Fördert gute Waldnutzung
FSC® C083411

Kolibri 360 GmbH
Ferdinand-Jühlke-Straße 7
99095 Erfurt, Deutschland
produktsicherheit@kolibri360.de